PUTTING THE BARN
BEFORE THE HOUSE

GREY OSTERUD

PUTTING THE BARN
BEFORE THE HOUSE

Women and Family Farming
in Early-Twentieth-Century New York

CORNELL UNIVERSITY PRESS   ITHACA AND LONDON

First published 2012 by Cornell University Press
First printing, Cornell Paperbacks, 2012
Printed in the United States of America

Library of Congress Cataloging-in-Publication Data

Osterud, Nancy Grey, 1948–
  Putting the barn before the house : women and family farming in early-twentieth-century New York / Grey Osterud.
     p. cm.
  Includes bibliographical references and index.
  ISBN 978-0-8014-5028-0 (cloth : alk. paper)
  ISBN 978-0-8014-7810-9 (pbk. : alk. paper)
1. Women in agriculture—New York (State)—Nanticoke Valley—History—20th century.   2. Rural women—New York (State)—Nanticoke Valley—History—20th century.   3. Family farms—New York (State)—Nanticoke Valley—History—20th century.   4. Farm life—New York (State)—Nanticoke Valley—History—20th century.   5. Nanticoke Valley (N.Y.)—Rural conditions.   I. Title.
  HD6077.2.U62N39   2012
  306.3'6150974775091734—dc23   2011041849

Cornell University Press strives to use environmentally responsible suppliers and materials to the fullest extent possible in the publishing of its books. Such materials include vegetable-based, low-VOC inks and acid-free papers that are recycled, totally chlorine-free, or partly composed of nonwood fibers. For further information, visit our website at www.cornellpress.cornell.edu.

Cloth printing   10 9 8 7 6 5 4 3 2 1
Paperback printing   10 9 8 7 6 5 4 3 2 1

Contents

List of Illustrations

Acknowledgments

The women and men who shared their recollections of living in the Nanticoke Valley cannot be thanked by name lest their identities be revealed, but this work was inspired by their remarkably honest and insightful reflections on the past. All are now deceased, but their loved ones may recognize them behind their pseudonyms.

The people who assisted me with this work but whose life stories do not appear in these pages can be thanked by name. Janet Bowers Bothwell, the curator of the Nanticoke Valley Historical Society, and Lawrence Bothwell, the Broome County historian, originally invited me to do research on the history of women in this rural community. Their legacy is embodied in the conservation of the historical landscape and in the museum and its collections. When I turned from examining the records of generations past to interviewing older women about their own lives, many people generously introduced me to their relatives, friends, and neighbors. Others graciously invited me to stay in their houses when I returned each summer from wherever I was teaching. Members of the Nanticoke Valley Historical Society have helped all along the way, most recently curator Sue Lisk, Nancy Berry, Alice Hopkins, and Sandy Rozek.

Over the years, many colleagues in women's history and agricultural history have commented helpfully on my analyses of these narratives and the transformation in rural society that occurred through the twentieth century. Those whose advice has been especially helpful include Hal Barron, Tom Dublin, Karen V. Hansen, Joan Jensen, Lu Ann Jones, the late Walter Meade, the late Mary Neth, Virginia Scheer, and the members of the Rural Women's Studies Association. Roy Christman, Deidre Crumbley, Karen V. Hansen, and two anonymous reviewers for Cornell University Press provided valuable feedback on the completed manuscript.

For the financial support that enabled me finally to finish this book, I am grateful to the Committee on Women Historians of the American Historical Association, which gave me the Catherine Prelinger Award in 2009–2010; I thank especially Eileen Boris and Nupur Chaudhuri for encouraging me to apply. Earlier stages of the research were supported by the National Endowment for the Humanities, San Jose State University, the American Council of Learned Societies, Lewis and Clark College, the New York Council for the Humanities, the New York State Council on the Arts, and the Historian-in-Residence Program of the New York State Historical Resource Center at Cornell University. I received valuable reference assistance from the staff of the Albert R. Mann Library at Cornell University; Delinda S. Buie, head of Special Collections at the University of Louisville; the staff of the New York State Library in Albany; and Jim Folts of the New York State Archives in Albany. I also thank the University of Louisville and the Nanticoke Valley Historical Society for allowing me to use photographs from their collections.

At Cornell University Press, Michael McGandy recognized that this book is a sequel to *Bonds of Community: The Lives of Farm Women in Nineteenth-Century New York,* which was published twenty-one years ago. Peter Agree has consistently supported my work from then until now. Karen Laun shepherded the manuscript through production. Kate Babbitt compiled the index and corrected the proofs. Bill Nelson designed the maps.

Portions of the introduction previously appeared in Grey Osterud, "Farm Crisis and Rural Revitalization in South-Central New York during the Early Twentieth Century," *Agricultural History* 84, no. 2 (Spring 2010): 141–65, © the Agricultural History Society, 2010. Portions of chapter two previously appeared in Grey Osterud, "Inheriting, Marrying, and Founding Farms: Women's Place on the Land," *Women's History Review* 20, no. 2 (April 2011): 265–81, © Taylor & Francis. Excerpts from Mary Neth, *Preserving the Family Farm: Women, Community, and the Foundations of Agribusiness in the Midwest, 1900–1940,* 32–33, © 1995 The Johns Hopkins University Press, are reprinted with the kind permission of The Johns Hopkins University Press.

Finally, as always, I thank my family and friends in New York, Ohio, Oregon, California, Massachusetts, Minnesota, and Utah for living with me as I lived imaginatively in the Nanticoke Valley.

TOWN OF
NANTICOKE

Nanticoke

N

Glen
Aubrey

Tiona

Allentown

Delano's
Corners

Bowers
Corners

TOWN OF MAINE

Mt. Ettrick

West
Chenango

King
Hill

Maine

Pitkin
Hill

East
Maine

Broughamtown

Nanticoke Creek

Bradley Creek

Bradley Creek

Choconut Creek

Union
Center

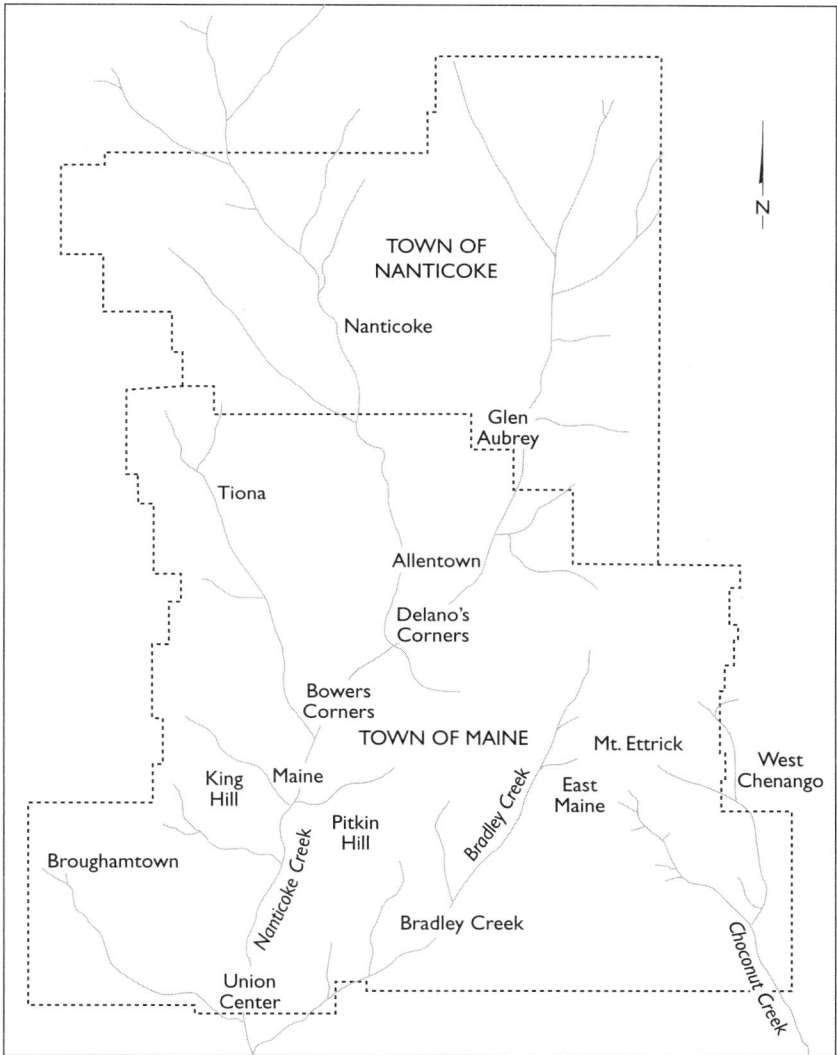

Map 1. Towns of Maine and Nanticoke, New York.

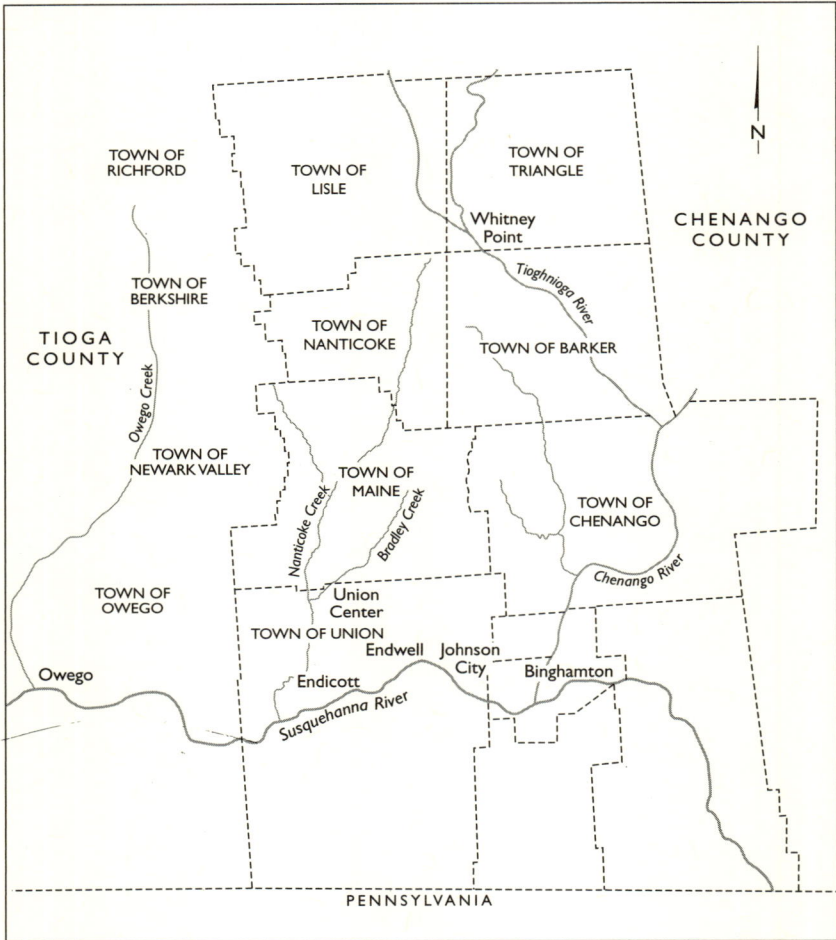

Map 2. Places adjacent to the Nanticoke Valley. Broome County is bounded by the broken lines.

PUTTING THE BARN
BEFORE THE HOUSE

Introduction

~

The Nanticoke Valley
in the Early Twentieth Century

People who drive through the Nanticoke Valley of south-central New York today find it difficult to imagine the intricate patchwork of farms that covered the countryside in the early twentieth century. The road following the Nanticoke Creek as it winds south from the upland towns of Nanticoke and Maine to join the Susquehanna River at Union passes scattered nineteenth-century farmhouses with dilapidated barns, Cape Cod–style houses with tidy flower gardens, and overgrown trailers surrounded by broken-down cars and rusting machinery. A few crossroads are marked by straggling hamlets, but none of the three villages boasts a grocery store.

Only one dairy farm remains in operation. At the northern end of the valley near Whitney Point, the Whittakers milk four hundred cows three times a day, use a computerized system to calibrate each cow's milk yield and nutritional needs, and grow feed and fodder on more than 1,000 acres of land. In following the recommendations of scientists at Cornell University, they represent contemporary agribusiness. At the southern end of the valley near the "Triple Cities" of Binghamton, Johnson City, and Endicott, the Wrights have given up dairying. They sold their land in the adjacent town to the Whittakers. On their farm, which has been in the family for 170 years, they raise vegetables, free-range hens, and grass-fed beef cattle, selling produce, eggs, and meat to urban and suburban customers much as the family did at the turn of the twentieth century. With a hint of humor leavening their seriousness, the Wrights say that their main crop is actually red-wing blackbirds, which nest in the rafters of their huge empty dairy barn.

Figure 1. Maine village, New York, ca. 1900. Looking east from King Hill toward Pollard Hill. Photographer unknown. *(Courtesy of the Nanticoke Valley Historical Society.)*

Most residents of the Nanticoke Valley do not cultivate the soil or tend livestock. The open landscape of a half century ago, with cornfields filling the valley floor, hay meadows on the gentle slopes, and pastures stretching over the rounded ridgelines, has disappeared under second-growth forest, with stands of neatly planted yew and pine alternating with thick scrub. Dark hemlocks have crept down from the ledges into stone-walled fields. Even the open vistas have vanished. Within living memory, the labors of generations of farming families have been erased from the land.

Nonetheless, new, market-oriented enterprises have sprung up in the Nanticoke Valley during the past few decades. Not all have survived, but gardens and greenhouses are thriving along some stretches of the creek road. The plot that Leigh Ames so carefully enriched to grow vegetables continues to flourish, and Black Angus cattle graze in their pastures, although the Ames's dairy barn is now used to repair cars. The Green brothers' fruit farm near West Chenango has become a "you-pick" operation and offers a restaurant, a gift shop featuring homemade jams and jellies, and a petting zoo. Those who first planted apple trees there would be startled, and then amused, to hear that a cow is now an exotic animal! Labor-intensive enterprises like the Wrights' depend on customers who are willing and able to pay a premium for quality. Their future is in doubt because the Triple Cities have been in decline since the post–World War II collapse of the shoe industry and the recent contraction of high-tech manufacturing.

Rural residents who now work elsewhere are aware of their community's past. Remarkably, the Nanticoke Valley Historical Society is the largest secular organization in the towns of Maine and Nanticoke, bringing together natives and newcomers to preserve and interpret their heritage. Being "from here" carries no special cachet. Not only does no merit attach to ancestry, but people deliberately refrain from making social class distinctions. Nobody pays much attention to the status of the jobs that others hold, whether they are executives, engineers, and professionals or work on the county road crew and in nursing homes. People care what others contribute to the locality, and the Volunteer Emergency Squad enjoys the most prestige. Those who have chosen to move to this community share many values with those who grew up here and decided to stay or return. All treasure the small scale of the built environment, with its modest houses and winding two-lane roads.

The Nanticoke Valley has been protected from development by being too far from the interstate highways that run between Binghamton and Elmira, Syracuse, Albany, New York City, and Scranton–Wilkes Barre, Pennsylvania. But its residents have also deliberately kept out interlopers that would have destroyed its character. Two decades ago they defended it against the U.S. Army Corps of Engineers, which planned to construct a flood-control system that locals deemed unnecessary and that would have required widening the creek road, demolishing historic buildings, and removing old trees. Although no historic districts have been created, restrictions have not been necessary to ensure that older structures are preserved. After conducting a survey of buildings and bridges, people put up plaques but decided against demarcating special districts. One longtime resident joked that nobody would want their house left out!

The impulse to record and preserve local history has deep roots in this community.[1] Unlike many small towns, Maine and Nanticoke did not suffer a century of historical amnesia between the late-nineteenth-century atlases, filled with biographies of white "pioneer settlers" and engravings of their imposing houses and barns, and the grassroots-oriented bicentennial celebration in 1976. Historical pageants were always popular. This awareness was complemented by a concern for the built environment. The Bowers family, who had been business associates of the Rockefellers, preserved several federal-period houses in Maine at the same time the Rockefellers were promoting the reconstruction of colonial Williamsburg. Clement G. Bowers, a botanist, knew that ecological relationships mattered. He preferred conserving old structures in situ to moving them to an open-air museum or building copies from scratch. A cluster of houses and shops dating from the late eighteenth to the late nineteenth centuries lends character to Bowers Corners. The family's influence prompted others to take good care of their historic homes as well.

Today, as in the past, the people of the Nanticoke Valley seek to live in harmony with their environment and sustain continuity with those who created the place they call home. Like those who went before them, they organize themselves to act cooperatively. The Nanticoke Valley Historical Society (NVHS) acquired and stabilized Pitcher's Mill, which used water power to grind grain into feed and sold buckwheat pancake flour; along with the adjacent Norton carriage shops, the mill shows how productive this countryside once was. The historical society also restored a one-room schoolhouse, which is open to school groups. The J. Ralph Ingalls School, which was built in neocolonial style in 1940, is now being rescued from decay and turned into a community center.

The first major NVHS project was a museum of local history. After acquiring a Victorian house, preserving its ornate parlor, and converting the rest of the structure into modern exhibition galleries and storage space for its collection of artifacts, photographs, and documents, the NVHS mounted an inaugural exhibit called, simply, "Nanticoke Valley History." Once it opened, Janet Bowers Bothwell, the cosmopolitan curator, realized that it should be subtitled "The Men's Story" because women appeared nowhere in the exhibition—except in a single photograph. Knowing that women must have played an integral role in local history and wondering what "The Women's Story" might be, she sought financial support and recruited a historian to research and document their experiences, voices, and perspectives for a complementary exhibition. That was thirty years ago, and I was that historian. When people saw old photographs of women raking hay and hauling heavy milk cans to the creamery, as well as sewing and conducting box suppers, alongside excerpts from diaries detailing everyday routines, they realized that the history of women was all around them, in the photographs and documents stored in their attics, in the configuration of their houses and farmsteads, and in the records of their churches and farm organizations.

Nanticoke Valley Women's Stories

So impressed was I by the rich resources for reconstructing this community's past and by the delightfully original elderly people who still lived there that I continued to do research and write about the Nanticoke Valley. In particular, I sought to understand what social conditions had enabled the women I met to lead lives that suited them so well. Nowhere else, in the historical record or the present, had I observed such marked variations in the gender division of labor. These women performed whatever set of tasks they personally preferred, and couples shared the work as they thought best. Many women had been full partners on farms, working along with their husband, sons, and daughters in the barn and fields. Some preferred the kitchen to the outdoors and others the reverse, but each made her own choices, at least to the extent that the weather allowed. Other women held full-time jobs off the farm but kept the books and did the taxes. Almost all had a say in farm family decision making. Nowhere outside of African American communities had I seen women who expected and received such respect from others.

In contrast to common stereotypes of farm women as downtrodden drudges, most of these women spoke as if they were the authors of their own lives. Nor were they isolated, as prevailing images of rural regions suggest. Indeed, people were in and out of one another's households with a frequency and degree of intimacy I found astonishing. Everyone know everyone else's business, including which men drank too much, which wives were at risk of being mistreated, and which children might be abused or neglected. Neighbors as well as relatives felt authorized to intervene to protect the vulnerable. Here privacy was nonexistent, but the safety net woven of kinship and neighboring was intact.

At the same time that I explored the history of this rural community, I realized that its social fabric was not merely an artifact of the past.[2] Although the culture of mutuality across lines of gender and generation and of local cooperation had deep roots, it had survived and thrived because a substantial number of newcomers had adopted and sustained it. The population of the Nanticoke Valley changed dramatically in the early twentieth century, as long-established families abandoned hill farms for urban employment and were replaced by immigrants fleeing the mines and mills. Ethnic and religious diversity replaced the homogeneity that had underlain unanimity of opinion. The rural economy was transformed from a mix of diversified farming and small industries that supported agriculture into a more stratified system, with some large-scale commercial farms and many small-scale, part-time farms. By 1930, almost every household sent at least one person to work for wages in the city. Yet these rural residents remained quite different from workers who lived in town. Despite the disparities among country people in terms of education and economic position, they collectively espoused egalitarianism and inclusiveness, practiced conspicuous restraint in consumption, and undertook projects collectively, particularly in agriculture and environmental preservation.

What was responsible for the remarkable degree of gender equality and neighborly cooperation that I discovered alive and well in the Nanticoke Valley? In this book, I present the answers I have found to that question. Some factors are structural: the economy was based on dairy farming, which demands the labor of all household members and requires flexibility in other work routines. Dairy farmers have formed producers' cooperatives, such as creameries, since the late nineteenth century, and collective action became even more vital during the consolidation of dairy processing and marketing in the early twentieth century. Some patterns are sociological: lifelong friendships were forged across lines of ethnicity and religion by children who attended one-room district schools, and a dense web of kinship was knitted together by marriages among young people who grew up in open-country neighborhoods. Rural social networks and formal organizations exhibited a striking capacity to incorporate newcomers from different backgrounds and mobilize them for common purposes.

Most important, the mutuality that existed within farm families was reinforced by cooperation among neighbors. As Mary Neth, a feminist historian of agriculture, put it in her germinal work, *Preserving the Family Farm,* "Promoting mutuality was a strategy that encouraged farm survival and improved the status of dependents within farm families. By emphasizing work flexibility, shared responsibilities, and mutual interests, farm people limited the conflicts created by the patriarchal structure of the family" and generated reciprocity within the community.[3] Until World War II, farmers' practice of "changing work" and their reliance on interdependence served as a viable alternative to capitalist agribusiness and provided the foundation for organizing. The values they held dear and their vision of a cooperative political economy sustained the farmers' movement. In this way, rural women's strategies of mutuality and farmers' practices of cooperation supported and enhanced one another.

It is impossible to re-create these socioeconomic conditions in the early twenty-first century; today they no longer exist even in the Nanticoke Valley. But it is possible to reconstruct the experiences of women who grew up in or entered this social world and to analyze their perspectives on gender and generational relations during what Hal Barron, a social historian, calls the second great transformation of rural society.[4] As I talked with older women about their ancestors, parents, and neighbors, I began to convince them that they, too, had stories worth telling. Most initially protested that they had nothing to say because they had done nothing unusual; they had led ordinary lives within a small circle of family and friends and had not participated in the major national events that they thought made history. I told them that, as a social historian, I was interested in common people's everyday experiences and their reflections on their own lives. Gradually they were persuaded to talk about themselves as well as about their elders.

I was blessed to be able to interview two dozen women who were born before the Great War (as the First World War was called before the second began). Some were the granddaughters of the women and men whose late-nineteenth-century diaries and letters I had read; these women were all Protestant, and most were of Yankee and Yorker descent. Others were the daughters of immigrants who had come to this rural locality in the early twentieth century, directly from nearby cities and indirectly from southern, central, or eastern Europe; many were Catholic, and a few were Orthodox. I spoke with women from affluent families and from impoverished ones, with those whose families led local organizations and those whose kin groups were socially isolated. Working my way along lines of personal connection and down the socioeconomic scale took time. Slowly, as residents learned over the years that I would never repeat what anyone had told me to anyone else and that I never judged anyone because of circumstances that were beyond her control, they spoke to me more frankly. The fact that I did not live in the community was crucial; people could confide their secret sorrows, recount their humiliations, and express their deepest doubts to me because they knew they would not meet me in the post office and have to face uncomfortable aspects of their personal past. It also helped that, although I was married, I was young enough to be these women's granddaughter.

The process of oral history interviewing was long term and collaborative, in keeping with feminist approaches that try to redress the imbalance between the narrator and the researcher. I not only listened to the stories these women had to tell but also attended to the ways they structured their narratives. In repeated interviews over several years, I probed their silences, explored the contradictions within their accounts, and invited them to consider the more problematic or troubling aspects of their past. Even more important, I shared interpretive authority, giving them transcriptions of their interviews and discussing the meaning of key turning points in their lives with them.[5] Taking women's politics seriously, I elicited their opinions on the rise of agribusiness and recent changes in women's lives. Their retrospective views, which were influenced by their present circumstances, have shaped the interpretations offered in this book.[6] Sometimes I saw things differently than they did, largely because, as an academic historian, I take a long-term social-historical perspective. In these cases, I have given both their viewpoint and my own.

Debating Rural Decline

Where I differed most profoundly from some of the elders with whom I spoke is in my assessment of the historical trajectory of the Nanticoke Valley itself. In retrospect, many lifelong residents regarded the transformation of the landscape as the result of economic and social forces outside the locality. In their eyes, a thriving rural community was disrupted by two world wars and subjected to capitalist consolidation as agribusiness superseded family farming. The fundamental changes that became painfully apparent after 1945 had, they recognized, begun earlier, as rural life was gradually eclipsed by urbanism. The decline they described was moral as well as economic; people had been seduced by the eight-hour workdays, household conveniences, and consumer pleasures of the city. Deserted first by their age-mates and then by their children, these men and women stayed on and patched together a living, only to watch the schools consolidate, churches merge, and village shops be converted into dwellings. Of the once-lively local social life, only dish-to-pass suppers in church basements continued, becoming a ritual whose familiarity was reassuring but a bit lacking in spice. To go to work, shop, or see a movie, people drove to the nearby city. In their view, the rural community had lost its integrity, surrendering its sturdy self-sufficiency and becoming indistinguishable from the urban society they deplored.

Although this perspective eloquently expresses longtime residents' sense that change came as a result of forces they could not control, it projects the causes of change onto the outside world rather than acknowledging their roots in local conditions. These stories of declension bore some resemblance to the early-twentieth-century reports of metropolitan observers who had worried about rural decline. Yet these critics articulated a very different analysis of the cause of rural problems. In their view, the countryside was characterized by economic decay and social stagnation. The flight of rural youth to the cities and the abandonment of farms, they contended, were prompted as much by cultural as economic deprivation.[7] Farmers left, businesses closed, participation in religious and civic activities declined, villages vanished, and the remaining rural residents were more isolated and backward than before. The remedy, according to these outsiders, lay in the adoption of urban social patterns, including a competitive, capital-intensive approach to agriculture; social relations that clearly differentiated people along the lines of age and gender; and more formal, large-scale organizations that linked people on the basis of interests rather than locality. Although this perspective recognized the indigenous roots of rural problems, it seriously underestimated country people's resistance to capitalist culture and their ability to adapt to changing conditions while maintaining their distinctive identity and values.

Ironically, both the rural jeremiad and the urban critique assumed that American farmers were passive in the face of economic and social change. Those who lamented their demise regarded farming families as helpless victims of the assaults of capitalism and metropolitan culture, while reformers saw them as benighted objects of ministration by benevolent outsiders. Both views were distorted by a false separation between rural and urban society. A more comprehensive perspective locates the causes of change not in urban imposition or rural decay but, rather, in the dynamic relationship between the countryside and the city. That interaction reshaped the rural political economy and country life, and farm people were active participants in socio-economic transformation. The process of adjustment that took place in the Nanticoke Valley between 1900 and 1945 shifted the relationship between the country and the city and enabled old and new residents to sustain the economic viability and distinctive identity of rural society.

As historians are well aware, people's recollections and interpretations of the past are shaped by their present circumstances as well as by their life experiences and deeply held values. The laments about the decline in rural culture that I heard from so many longtime residents of the Nanticoke Valley registered the dramatic changes that had taken place since World War II and tended to elide the more recent past with the subtler shifts that had occurred during the first half of the twentieth century. I have linked their rich accounts of the past to the documentary evidence to discern group-level patterns of social interaction.

The rest of this introduction traces fundamental changes in the rural economy, especially the trend toward combining farming with urban wage-earning, and examines emergent patterns in rural society, especially the relationships between natives and newcomers that developed as many families departed and immigrants arrived in open-country neighborhoods. A complex process of social reorganization enabled people to adapt to change and accommodate difference without fracturing the sometimes-fragile consensus on which rural culture depended.

At the turn of the twentieth century, the Nanticoke Valley was relatively homogeneous and socially integrated yet in the midst of a long-term process of demographic decline. The population had fallen steadily since its peak in 1880; in 1900, there were 2,200 people living in 627 households in the towns of Maine and Nanticoke, down from 3,128 two decades before. The population continued to fall until 1920, when there were just 1,804 people in 534 households. On average, rural households were quite small, with 3.5 people each; in one in five households, a person lived alone. Nanticoke Valley families no longer resembled their counterparts of a half century before, when parents raised comparatively large numbers of children and worried about how to help them all get established on the land. The population was aging, with relatively few children and working-age adults; just one-third of residents was under twenty years of age, and one-sixth was sixty-five or older.[8]

The causes of this demographic decline were obvious to everyone. The supply of undeveloped land suitable for agriculture was exhausted, and most farms were too small to subdivide. So, although one son or, lacking sons, a daughter might look forward to inheriting the family farm, the other children had to find ways of making a living outside of agriculture. The countryside had deindustrialized during the last two decades of the nineteenth century because the small mills and manufactories could not compete with larger, more mechanized processors and producers in the cities. Local sawmills continued to do custom work for farmers harvesting trees from their woodlots, but most houses and barns used standardized millwork from steam-powered plants that was shipped in by train. Small water-powered furniture manufactories closed as cheap mass-produced chairs, tables, bureaus, and parlor suites became available in city showrooms and mail-order catalogues. Blacksmiths shifted from making to repairing tools. Carriage shops no longer made wagons and wheels from scratch, but assembled and painted vehicles that were mass-produced elsewhere.[9] When people without land found it almost impossible to earn a living in the country, they had to leave. Few had the capital required to move to the Midwest to farm, as some of their parents' and grandparents' siblings had done. Homestead land was no longer available, except in the least fertile and most forbidding regions. Buying enough prairie acreage and the horse-drawn machinery required to cultivate it was prohibitively expensive; even selling a successful farm in Broome County would not bring the amount of money it required. Most young people who grew up without the prospect of inheriting an operating farm migrated to urban areas to work for wages. Even if they moved only 15 miles over the hills to the commercial city of Binghamton or settled in the adjacent manufacturing villages of Lestershire (later called Johnson City) and Endicott, they entered a different social world.[10]

Those who remained in the Nanticoke Valley adjusted their expectations to local circumstances. Keeping the farm in the family—laboring to improve the land, struggling to pay the taxes and at least the interest on the mortgage, and passing on a viable operation to a grown child—was as much as most families could hope for and more than many could attain. Families expanded their commercial dairy operations, relying on the relatively small but steady income to maintain their farms while continuing diversified subsistence production and small-scale, market-oriented "sidelines" to sustain themselves. Their horizons were limited because the accumulation of wealth was beyond their reach. But most found a certain satisfaction in upholding their place in local society and valued the dense web of relationships that connected them with relatives and neighbors.

The process of economic and demographic contraction that occurred during the late nineteenth century actually facilitated the formation of stable social networks. As Hal Barron pointed out in his study of nineteenth-century Vermont, population stagnation, or even decline, did not disrupt the lives of "those who stayed behind"; instead, continuous outmigration allowed rural residents to maintain their way of life in spite of economic contraction. Most people had grown up in the immediate vicinity; they were bound together not only by lifelong association but also by ties of kinship and friendship several generations deep. The rural population was relatively homogeneous, and public life appeared harmonious. The social and political conflicts that had accompanied economic expansion, especially in a diverse and dynamic population, were either resolved or removed from open debate; consensus and the avoidance of conflict characterized community life. Social stability accompanied rural depopulation.[11]

By the Great War, however, Nanticoke Valley residents were becoming worried about the massive outmigration from upland farms to the urban areas of Broome County. The exodus of young people had increased quantitatively to the point that it differed qualitatively from the previous pattern. Farmers now faced a genuine shortage of labor. The high wages offered by factory jobs, coupled with the rapid inflation of prices for consumer goods, drew unprecedented numbers of young men and women into the cities. No longer could farmers hire their neighbors' maturing sons to help with haying and the harvest and their growing daughters to work in the dairy. Indeed, few could keep their own sons and daughters at home. Some families, especially those in upland areas, left the land altogether. Farms were rented out or left vacant, and fields grew up to brush. It was apparent to the remaining farmers that migration to urban areas no longer served as a safety valve for a surplus population but was undermining agriculture and the rural society that depended on the land.

People who stayed in the Nanticoke Valley interpreted this crisis as one of family succession and neighborhood stability as well as economics. The sons and daughters of farming families preferred the regular wages and limited hours of factory jobs to the unremitting and often unrewarding toil on the land. The amenities of urban life, ranging from movie theaters to indoor plumbing, were more attractive than the isolation and inconvenience of country houses. Youths who attended high school and found jobs in the city seldom returned; they established themselves in nonagricultural occupations, married, and aspired to purchase modest homes in urban neighborhoods. Even those young men who initially stayed on the farm, working with their parents and hoping to inherit the land eventually, often became discouraged and departed because the prospects seemed bleak and the delay intolerable. Neither generation had the capital required to expand the enterprise so it would be able to support two families. Most aging parents could not afford to retire without selling the land, and most adult children could not afford to buy them out. Many couples continued their customary pattern of farming as long as they were physically able to do the work and did not expect their children to succeed them. When they died or were forced by ill health to quit farming, the land was sold or left vacant. The failure to secure intergenerational succession was a serious matter; those who identified their families with the land they had inherited felt that they had broken faith with their forebears. Yet they did not blame their children for leaving. Rather, they understood that their own economic position made it at best difficult, and at worst impossible, for them to fulfill their aspirations.

As the farm population aged and declined, the character of rural neighborhoods began to change as well. Distances between neighbors increased, both physically and socially. In upland areas whole hillsides reverted from pasture to woodland, and inhabited farmhouses stood relatively far apart. The dispersed pattern and declining density of settlement made daily visiting more difficult. Neighbors were less likely to be close relatives than they had been previously, so intimate forms of contact and substantial material aid became less common. However, the norm of neighborly cooperation remained firmly ensconced in the rural value system.

Reforming the Agricultural Economy from the Grassroots

The demographic and economic history of the Nanticoke Valley during the late nineteenth and early twentieth centuries resembles that of many other rural areas in the northeastern United States. By 1910, the exodus of young people and the abandonment of farms had become a matter of national concern. Responding to an unsettling sense that the iconic American farmer was disappearing, the Country Life Commission appointed by President

Theodore Roosevelt focused its attention on rural social problems. Its report looked to country people to provide a bulwark against the changes demanded by the urban, largely immigrant working class, which was expanding rapidly and mobilizing politically during this period. The problem, as these metropolitan observers saw it, was that white Americans of native parentage whose attachment to the land made them the best popular defenders of the rights of property were abandoning the virtues of rural life and defecting to the cities, where they might be corrupted by the "foreign element" of political radicalism.[12]

The solution suggested by agricultural economists and rural sociologists—transforming farming to make it more profitable and reforming rural communities to make them more attractive—was politically conservative, but it had little or nothing to do with preserving a threatened way of life. The rural idyll these reformers imagined in the past had never existed. Across the country, farm people had sought to protect and advance their economic and social interests by adhering to noncapitalist or anti-capitalist ideas and practices that valorized producers over "parasites," resisting their subordination to profit-minded shippers and processors, and refusing the blandishments of political leaders who told them that what was good for the commodity market was good for farmers. The Country Life Movement implicitly accepted capitalist control of the agricultural economy and explicitly endorsed the trend toward commercialization and specialization that was leading to the consolidation of farms and the displacement of families from the land. Experts' advice to farmers, which centered on the application of cost-profit analysis to farm operations and the adoption of capital-intensive farming techniques, was designed to make individual farms more competitive in the marketplace, not to restructure the market or change the relations of power among producers, shippers, processors, and mercantile firms. Equally important, it ignored and, in practice, would have eroded the interdependent relationships among farming families. The Country Life Movement attempted, with some success, to shift the terms of public debate away from the Populist agenda and focus attention on the problems within rural society rather than on the unequal economic relationship between farmers and capitalist agribusiness.[13]

Broome County farmers rejected this diagnosis of their difficulties and prescription for change. Instead of adopting capitalistic solutions to their predicament, they took over the countywide agricultural improvement organization formed by Binghamton's business leaders, renamed it the Farm Bureau, and transformed its mission, focusing on raising farmers' incomes rather than on producing cheap food for urban dwellers. They eagerly adopted the scientific methods offered by experts at Cornell University's New York State Agricultural Experiment Station but rejected the business models its farm management advisors recommended. Instead, they promoted producers' cooperatives—associations that sold their members' produce, as well as purchasing fertilizer and feed, and used their size and strength to drive better bargains with processors, suppliers, and shippers. Building on the strong local tradition of the Grange, which had founded the cooperative creameries that were still in operation in the Nanticoke Valley at the turn of the twentieth century, they joined the Dairymen's League en masse in 1919 to negotiate higher prices for the fluid milk they sold to dealers in New York City.[14]

Farmers' customary patterns of cooperative work and the habits of mutuality that pervaded open-country neighborhoods provided a foundation for collective action. Nanticoke Valley farmers established groups in Nanticoke, Glen Aubrey, Maine, and Union Center that included hill farmers with marginal land as well as valley farmers with somewhat less stubborn soil. Local solidarity proved crucial in the 1919 milk strike and the subsequent campaign to secure contracts that equalized the return to large fluid milk producers and smaller producers who sold cream to condensaries; neighbors pressured neighbors to withhold their milk and then to sign pooling contracts.

Older men and women whose parents had been active in the Grange, the Farm Bureau, and the Dairymen's League and who remembered these struggles from their youth expressed pride in the fact that in the Nanticoke Valley, unlike other places nearby, there had been no incidents in which striking farmers seized and dumped milk cans that nonstriking farmers were taking to shipping stations because nobody even tried to ship milk.[15] Carrie Northfield of Nanticoke said that all of their neighbors observed the strike, although there were violent incidents just to the north near Whitney Point. Ralph Young, who farmed with his father and brother near Union Center, explained, "Sure, there were some farmers who thought the strike could never succeed and who were afraid to break their contracts with the processors. Mostly, they had larger herds. But they also depended on their neighbors at haying time and on hired hands for help with chores, so they could be persuaded. Not violently, but...they could be persuaded."

Gender integration was as marked a characteristic of local farm organizations as their neighborhood base. Although the state-level leaders of the Farm Bureau and Dairymen's League thought of farmers in the masculine gender, grassroots activists did not. The Grange tradition of family membership was more in accordance with local gender-integrated customs of farming and organizational life. Nanticoke Valley residents acted as if the Farm Bureau were a mixed-gender group; husbands and wives joined together and shared leadership responsibilities. Not even the establishment of the Home Bureau as a distinct department designed to serve the needs of women led to significant gender segregation. Broome County adopted a coordinate structure that provided for joint decision making and common activities, officially becoming the Farm and Home Bureau in 1920. In April 1923, the *Broome County Farm and Home Bureau News* reported on a dinner held by the Maine Dairymen's League to discuss the development, problems, and projects of the organization: "One of the interesting things about the meeting was the presence of the women folks. They took part in some of the discussion too, which shows that they are interested in the problems which confront the dairymen today. This is as it should be, and there is room for many more such gatherings where both men and women talk over their problems frankly together." Members would not have appreciated the editor's condescending tone, for in their experience both women and men were integral to dairying. The tradition of flexibility in farm work and shared family decision making, as well as joint participation in neighborhood and organizational life, permeated and reshaped the activities of supposedly masculine groups such as the Dairymen's League and ostensibly gender-segregated groups such as the Farm and Home bureaus. In the process, women as well as men were enlisted in securing the unity of action on which producers' cooperation depended.[16]

The people of the Nanticoke Valley responded to the farm crisis of the early twentieth century by sustaining cooperatives and undertaking new forms of collective action. Customary gender-integrated modes of neighborly association provided the base for powerful economic and political organizations.

Forming New Connections between the Countryside and the City

People not only responded to economic crisis with concerted action but also adjusted gradually, although sometimes grudgingly, to changing conditions. New connections between the Nanticoke Valley and the industrial centers of Broome County made rural revitalization possible.

From 1917 on, rural residents could work in the city and urban workers could live in the country. At the height of the war mobilization, a private company began to operate a twice-daily "workers' bus" on the newly macadamized road from the Nanticoke Valley to Union. Some people went to the boot and shoe factories in Endicott and Johnson City, and others took the Triple Cities streetcar to jobs in Binghamton. As soon as civilian vehicles were available, farm families purchased automobiles and trucks, which they used not only to carry goods to market but also to commute to work. At the same time, urban workers began moving to the countryside. Many of the newcomers were immigrants from rural regions in central and eastern Europe who had held on to their agricultural aspirations through years of toil in railroad construction, in the anthracite mines and silk mills of northeastern Pennsylvania, and in the tanneries and boot and shoe factories in Endicott and Johnson City. Poles, Slovaks, Ukrainians, and Armenians[17] invested their precious savings in run-down farms. While some members of their families continued to work in the city, the others devoted themselves to agriculture. When local youths no longer had to leave the countryside to find employment and new families began moving in, the decline in population was arrested and rapidly reversed.

Between the First and Second World Wars, the Nanticoke Valley enjoyed modest growth. The population of the town of Maine, which was closest to the urban areas along the Susquehanna River, rose from a low of 1,360 in 1920 to 2,076 by 1940, while that of the town of Nanticoke rose from 444 in 1920 to 546 in 1940. In 1940, the 2,622 residents of the Nanticoke Valley included many more children; almost one-third of the population was younger than fifteen. The proportion of people of working age was also much larger; less than one-tenth was sixty-five or older. The newcomers from the city included many young couples who brought up their families in the countryside and contributed mightily to the demographic renewal of the Nanticoke Valley.[18]

The rural economy was supported, rather than undermined, by its close reciprocal connections with the urban economy. Families combined wage labor, subsistence production, small-scale market gardening and poultry raising, and commercial dairying in a variety of ways. They could shift the balance of their efforts over time, as the composition of their household or the relative advantages of different activities changed. This flexibility was especially important during the depression of farm commodity prices in the 1920s and the collapse of the urban labor market in the 1930s. Families invested their savings from wage labor and from selling farm products in purchasing land and expanding farm operations during the 1920s, and they relied on their subsistence production and commercial operations during the 1930s. Flexibility was central to families' survival strategies, and economic diversity sustained the locality.

Some farmers continually expanded the scale of their operations to maintain the profitability of their enterprises. They kept large dairy herds and sold fluid milk, which required them to pour a cement floor in their barn and construct a holding tank. A few raised poultry to supply other farms with chicks as well as sending eggs and chickens to market; others had commercial apple orchards. These farmers readily adopted new techniques and machinery once their practical benefits had been demonstrated, provided that the innovations would pay for themselves in a few years. They led the new farm organizations and sponsored local meetings for the discussion of farm problems. Their economic reliance on marketing milk and purchasing feed, coupled with their custom of changing work with their less prosperous neighbors, ensured that they reached out to others.

Most farmers operated on a more limited scale. They usually relied on time-tested agricultural methods because they had little capital to invest in improvements and were profoundly risk-averse. These families produced a wide range of marketable commodities, as well as providing for their own subsistence. Many had small dairy herds, which they milked only in the summer to avoid having to buy much feed. Some sold their milk to the creameries, and others made high-quality butter at home. They kept chickens, raised potatoes, and grew vegetables such as cabbage. Often they marketed their produce themselves, either in the public markets of Endicott and Johnson City or on regular routes through urban neighborhoods. People who had moved to the Nanticoke Valley from the city or sent a family member to work there every day were especially likely to sell directly to consumers because they had personal connections with particular neighborhoods or workplaces. Although these small operations appeared inefficient to outside experts, they provided much-needed cash income as well as a secure subsistence.

Many of these families aspired to farm on a larger scale, but relatively few succeeded in making the transition to specialized dairying. The prices of farm products were too low, and the capital costs of large-scale farming too high, to enable them to expand simply by reinvesting their profits. Native sons and a few daughters who inherited land and livestock relatively young, or who shared labor and machinery with their parents and siblings while they built up their dairy herds, might manage it—if they were lucky. Some rented or bought land from an elderly widowed relative, paying for it over the years in cash and kind. Everybody knew better than to take the risk of going deeply into debt, even if they could find a banker foolish enough to loan them money. Some determined families might subsidize the expansion of their farm operations with their earnings from off-farm jobs, provided that they were willing to forgo consumption and work a double day. But the majority continued to farm on a small scale, supplementing their farm income with wage labor.

For a significant minority of Nanticoke Valley families, in fact, farming supplemented wage-earning rather than vice versa. In 1925, 40 percent of all adult males in the community reported nonagricultural occupations on the New York State census, twice the proportion that had listed nonagricultural occupations in 1900. Over time, the distinction between farmers and wage workers became blurred. In 1940, 40 percent of farm operators in Broome County reported on the federal agricultural census that they had been employed off the farm during the previous year, averaging 187 days of paid off-farm labor in 1939. A similar combination of agriculture and industry had characterized the local economy in the mid-nineteenth century. Although farms and factories were now located some distance apart, the automobile enabled Nanticoke Valley residents to be part-time, small-scale farmers and casual laborers or permanent industrial workers. Their dual occupations complemented one another. Although they sometimes regretted the enormous amount of labor they performed and lamented their lack of control over either wage rates or farm commodity prices, they also prided themselves on their relative independence from bosses and milk dealers. When low commodity prices jeopardized their fellow farmers, they were relieved that they received a guaranteed return for their labor. When unemployment threatened their fellow workers, they were grateful that they would always have a place to live and could grow their own food.[19]

When the adult sons and daughters of farm families were able to live nearby and commute to work, intergenerational ties were retained. They sometimes rented or purchased a farmhouse or village residence and raised their children within daily visiting distance of their parents. In other cases, the younger generation purchased the family farmstead when the older one gave up farming. Aging couples sold their livestock and implements to other farmers to obtain a nest egg and received monthly payments from their non-farming children who had purchased the house and land, an arrangement made possible by the wages that the younger generation earned. Usually the two households cooperated in childcare and household repairs. If the older generation continued to farm, grown children also exchanged labor, cash, and consumer goods for farm produce. Proximity mitigated the most serious consequences of the failure of farm succession. Older people could rely on their married children for care in infirmity without having to forfeit their independence while the younger generation could provide supportive services for aging parents without having to take them into their households. Nanticoke Valley families had long struck a delicate balance of independence and interdependence between the generations, but at the turn of the twentieth century separation or dependence had seemed to be the only alternatives. The combination of urban wage-earning and rural residence enabled some families to sustain continuous, mutually beneficial ties between the separate households of aging parents and married children during the interwar period.

Revitalizing the Rural Community

The decline in neighborhood cohesion that had begun during the process of depopulation was not entirely reversed by the resumption of growth. The newcomers who purchased vacant farms were different, in both class and culture, from the farm families who remained. They were urban and working class, even though many of their families had been pushed off the land a generation before. Most were immigrants or the children of immigrants; many were Catholic. To native-born people of native parentage, they were all "foreign." The fact that some had been born or grown up in the United States made them no less strange. The ethnic neighborhoods on the North Side of Endicott and in the First Ward of Binghamton, from which these families came, seemed to longtime residents of the Nanticoke Valley as socially distant as the European countries the newcomers' parents had left. The fervent Americanization campaigns of the post–World War I period tended to exacerbate the antagonism toward ethnic minorities.

Relationships between natives and newcomers varied, depending on the class and cultural composition of particular neighborhoods. Immigrant families who moved into neighborhoods inhabited primarily by long-established families often remained isolated. Although the men on adjacent farms assisted one another, these families did not socialize as often as they did with people with whom they had longer associations. Natives some-times explained the decline in informal visiting and local get-togethers in terms of the changing ethnic composition of their neighborhoods, presum-ing that people who differed culturally would be socially distant as well. Newcomers shared their labor and leisure with their extended families and ethnic networks, which were still centered in the city. In some places, natives and newcomers inhabited separate social networks even though they lived next door. This pattern, however, did not prevail throughout the Nanticoke Valley because fewer immigrants moved into stable neighborhoods than into impoverished ones.

Some neighborhoods became populated mostly by newcomers. Upland areas that had largely been abandoned, such as Pitkin Hill, were resettled by urban workers in search of cheap land. Most of the new residents were im-migrants or the children of immigrants. But ethnic segregation was less pro-nounced on Pitkin Hill than in the urban neighborhoods from which these families had come; rural folks forged ties with one another through neigh-borly cooperation and informal socializing. The development of a neigh-borhood that was identified as "foreign" and "Catholic" was greeted with hostility by some native-born people, especially those who lived in more homogenous villages in the valley. The Ku Klux Klan (KKK), which led a virulent anti-Catholic, anti-immigrant campaign in the North during the mid-1920s, organized a chapter in Union Center, the predominantly native-born, Protestant village below Pitkin Hill.

Relations among natives and newcomers were closer and more harmoni-ous in neighborhoods characterized by class homogeneity as well as cultural diversity. Some upland neighborhoods, such as Tiona, were inhabited by a mix of lifelong residents and recent arrivals, small farmers and wage work-ers. Significantly, these social categories cross-cut one another; farmers and urban workers were represented among both natives and newcomers, and many families engaged in both types of labor. The commonalities of people's situation as they struggled to piece together a living from the wages of their uncertain and unrewarding jobs and the produce of their gardens, poultry flocks, and small dairies seemed more important than their differences. Cus-toms of cooperative work were reinforced by their lack of farm machinery. As the children attended the district school, they forged ties that soon drew their parents together. Mutual aid in crisis and shared recreation knit natives and newcomers into an integrated local social network.

Advocates of cooperation took strong exception to the notion that ethnic and religious differences constituted fundamental divisions among farm people; in their eyes, producers had to stand together against the processors and merchants who profited unfairly from their labor. The Grange, Dairymen's League, and Grange League Federation made deliberate efforts to involve immigrant farmers. Their approach was ideologically based as well as strategic, and it had deep roots in customary practices of neighboring. Although county-level leaders were most often native-born Protestants, local membership lists are peppered with Slavic, Polish, Russian, and Armenian names. George Young, a longtime leader in the Union Center Grange, referred to immigrant farmers as "new Americans" and insisted on recruiting them. He and his wife visited with immigrants who lived on nearby farms after their children got acquainted at school; he hired the boys to work alongside his sons during the summer while the women traded plant cuttings and newly hatched chicks and did their canning cooperatively. Soon these families were attending Grange picnics together.

Inclusiveness was a conscious, time-honored practice. In 1921, the Union Center Grange built a community hall that hosted Dairymen's League meetings and farmers institutes that were open to members and nonmembers alike. The Home Bureau conducted social events at the hall to benefit local causes; in 1921, for example, it held a clam chowder supper to raise money to buy playground equipment for the school. A photograph that appeared in the September 1921 issue of the *Broome County Farm and Home Bureau News* carried the caption: "Every community should have some center where all may meet regardless of race, creed, or worldly goods." This declaration was an explicit repudiation of the anti-immigrant, anti-Catholic stance taken by the KKK chapter in the village. Union Center held a Community Day picnic in 1921; a Community Council composed of representatives from all local organizations was established by 1927.[20]

As anyone who is familiar with interactions among rural residents is well aware, "the community" is not a natural formation inscribed in the landscape but a social and ideological construction deployed for specific purposes. Like "the nation," community is always more imagined than objectified.[21] Rather than a bounded polity, such as the two towns of Maine and Nanticoke, it refers to people's social networks and their shared practices and beliefs, as well as to their understandings of people and places that are outside those networks. Over time, who belongs and who does not shifts, as do the connections that constitute community. When Nanticoke Valley people spoke of the community in the singular, they emphasized the value of social solidarity in support of what they defined as the common good and in contrast to what they saw as others' self-interested actions. Some people—most often those who had spent time in cities, but also those whose extended families had a reputation they did not want applied to themselves—found the emphasis on conformity stifling and the assumptions made about others troubling. The rural populace was always more divided by class and culture than the reigning ideology acknowledged, and anxious avoidance of open conflict was a pervasive undercurrent in social life.

Some of the new forms of social organization that Nanticoke Valley residents adopted during the 1920s were recommended by the Country Life Movement as ways of revitalizing local institutions while preserving the moral values of rural life. But Nanticoke Valley leaders were more concerned that rural society might be disrupted by conflicts arising from the growing class and cultural differences within the population. Their response to this problem was the opposite of what reformers recommended; rather than consolidating local institutions, they ensured they were tied together by informal and inclusive customs of sociability and cooperation.

The community halls that were built in the Nanticoke Valley during the 1920s were designed to serve as central meeting places. An array of local groups not only held open meetings there but put on Home Talent plays, convened community sings, and showed movies. Most events mixed education and recreation with fund-raising. Local norms required everyone to support these events, regardless of their organizational loyalties. Moving these gatherings to a hall that belonged to everyone made them even more inclusive. But, as Nanticoke Valley residents soon discovered, it also created grounds for discord that were better avoided.

A photograph of the Glen Aubrey Community Hall illustrated an article by Dwight Sanderson, professor of rural sociology at Cornell University's College of Agriculture, published in the *New York State Farm Bureau News* in September 1921. The story that accompanied this article in the Broome County edition told how residents of the town of Nanticoke had converted the abandoned Christian Church in the hamlet and set up a board of directors with representatives from each of the groups that used its facilities. This governance model, although recommended by rural sociologists as a way of promoting unity, was vulnerable to conflict. The Glen Aubrey board reached an impasse over the question of which activities would be allowed in a building that had formerly been dedicated to worship, and the hall closed in the late 1920s. As Carrie Northfield explained the situation, people from different churches could never agree on whether it was acceptable or immoral to dance or play cards there, or even on whether the building was now a secular structure. Community halls in the other villages proved more sustainable. Like the Grange hall in Union Center and the Methodist Church hall in West Chenango, they were owned and operated by specific organizations but served many groups.[22]

The proliferation of organizations that rural sociologists criticized as a waste of scarce resources and a sign of social disintegration allowed rural residents to accommodate diversity while avoiding controversy. The Country Life Movement recommended the consolidation of churches so they could support a professional ministry and an active parish life in spite of rural depopulation. What the outside experts failed to understand was that separate organizations enabled distinct groups to feel that they controlled their own affairs, while cooperation ensured the pooling of resources and broad participation. The history of the churches in Maine village illustrates this dynamic. The Congregational, Methodist, and Baptist churches held separate services each Sunday morning, but a union service in the evening rotated among the three churches and was conducted by each minister in turn, with members of different denominations mixing in the pews. After the Methodists' church in the village burned, the Congregationalists invited them to form a Federated Church in 1929. Modifying the advice of a Cornell sociologist, they decided to maintain the formal distinction between the two denominations. They believed that this arrangement would enable them to avoid conflict over matters of theology and ministerial selection, and it has worked well ever since.[23]

In Union Center, Glen Aubrey, and Nanticoke, too, the social events put on by the Ladies Aid Society of any one church were attended by people from the other denominations. Religious exclusiveness was regarded as petty and potentially divisive. Catholic children were encouraged to attend Protestant social functions with their schoolmates, and anti-Catholic sentiment sometimes took the form of statements that Catholics were being disloyal to the community by going elsewhere to church. Eventually a Roman Catholic parish was established in the Nanticoke Valley. Not only did Catholics have to win acceptance from Protestants, but Catholics who came from a variety of ethnic traditions had to work out their liturgical differences. The catalyst for the formation of the new parish was a survey that revealed that a significant proportion of children were receiving no religious instruction. Evidently, their parents still identified as Catholics but no longer traveled into the city on Sunday. Residents thought of themselves as united by their commitment to religious observance and education. In the face of a threat to that shared value, they accepted a greater degree of religious diversity. The new parish purchased the old hotel in the center of Maine village, and Protestants as well as Catholics contributed to the renovation of the building. The Most Holy Rosary Catholic Church took its place beside the Federated and Baptist churches in 1942.

The relationship between the villages and open-country neighborhoods was carefully renegotiated during the 1920s and 1930s. Although village commerce had suffered as a result of urban competition, the disappearance of hamlets and the advent of the automobile maintained significant levels of business and social activity. Leaders were aware of the dangers this change posed as well as the opportunities it offered. Differences between village and open-country residents were still substantial, and any attempt on the part of villagers to dominate the others was likely to create antagonism. Town leaders, while using the villages as central meeting places, were careful to ensure the representation of open-country residents and consideration of their interests. The skill with which they handled village-countryside relations is evident in the gradual process of school centralization in the town of Maine. From the Country Life Movement on, reformers advocated a massive shift from locally controlled district schools to graded schools in central places, arguing that they offered superior facilities and expert instruction at a lower cost. Through the 1920s and early 1930s, in spite of state enabling legislation and considerable financial pressure to consolidate, Maine and Nanticoke maintained their one-room schools. Parents could send their children to the larger schools if they chose to do so, and rural districts could voluntarily merge with the village districts. But villagers refrained from appearing to pressure open-country residents to give up their schools, which were often the sole remaining centers of neighborhood identity and sociability. When school centralization finally came to Maine in 1938, all nine rural districts and the two village districts agreed on the construction of new buildings in Maine and Union Center to serve the entire town. Through careful management, town leaders ensured that the decline of open-country neighborhoods and the increasing diversity of the rural population actually increased social cohesion.[24] In consolidating schools and federating churches, Nanticoke Valley residents took experts' advice when it seemed useful but reshaped outsiders' recommendations to serve their own purposes.

The Nanticoke Valley retained its rural identity and its integrity as the center of social life for its residents through a creative combination of old and new forms of social organization. Its proximity to the urban area of Broome County provided the basis for its relative economic stability and renewed demographic growth. Those changes, in turn, supported the continued vitality of its social and cultural life. The Nanticoke Valley was just distant enough, and different enough, from the city to avoid becoming absorbed into or entirely dependent on urban culture. The influx of immigrants and the increase in class distinctions challenged the rural social order, but did not fragment it; local responses to change revitalized rural social life. That organizing process sustained customary links between country people and created new connections among them. As homogeneity and decline gave way to diversity and growth, the revitalization of rural society reaffirmed long-standing practices of solidarity and inclusiveness.

Listening to Nanticoke Valley Voices

We begin by exploring what women's autobiographical narratives reveal about their connections to the land on which they lived and worked. Part I considers the dilemma that farm families, whose resources were necessarily limited, constantly faced: Should they place a higher priority on raising their standard of living, epitomized by their house, or on increasing their productive property, represented by the barn? Women were the pivotal figures in this gendered economic debate. Chapter 1 juxtaposes a famous short story, "The Revolt of 'Mother,'" with the narrative of an impoverished woman who farmed in the Nanticoke Valley. Chapter 2 traces the paths through which women came to live and work on family farms—inheriting land from their parents, marrying an inheriting son, or founding a farm in partnership with their husband—as well as those who were displaced from the Nanticoke Valley.

Part II considers the ways in which farming families in the Nanticoke Valley and Broome County experienced and adapted to economic change. Part III probes the interconnections between these economic shifts and relationships between women and men. In American society more generally, the gender division of labor and decision-making power were closely intertwined. What supported the remarkably flexible, gender-integrated, and egalitarian practices that prevailed in this place and time? How did increases in the scale and specialization of farming affect women? How did families make decisions about working on the land and in the factory, and what were the consequences of their varied arrangements? Part IV highlights women's active participation in movements that resisted the rise of capitalist agribusiness and the marginalization of women's work that was often explicit in the recommendations of agricultural and home economics advisors. Although rural women were constrained by deeply rooted societal patterns of male dominance and by poverty, they were not merely reactive to forces beyond their control. Rather, they were authors of their own lives and agents of change in the economy and polity.

The stories I heard about life in the Nanticoke Valley during the early twentieth century centered on resourceful families putting together a living in a variety of ways and enjoying the flexibility in gender norms that local custom accorded them. A lively rural culture was knit together by resilient patterns of cooperative action and valued productive labor over possessions. What today might be called quality of life was rooted in farmers' relative independence from the dictates of the capitalist market. As a large-scale dairy farmer explained to me, "We can afford to keep the best for ourselves. In the past, farmers in straitened circumstances went without butter, eggs, and fresh greens and ate only the fly-specked apples in order to sell everything they produced. We consume the best of what we grow because it's healthy and delicious. After all, we grow people, not cows. The milk checks support our family and the help we hire; we don't work for the milk company."[25]

In nurturing people, farm families generated cooperation across lines of gender and generation, sustained customs of mutual support among neighbors, and resisted the inroads of commodity culture. Urban dwellers who hear about this place today often see it as a model for community-based sustainable agriculture and environmental conservation. To me, the Nanticoke Valley is not a utopia—a word that, after all, means "no place." It is a particular place with a history of its own. But the society re-presented here to readers who live in a very different world points to the possibility of people acting collectively to create and sustain an alternative way of life at the same time that they adapt to changing economic circumstances. The relative gender equality, class and ethnic inclusiveness, and modes of cooperative action that still prevail in this locality testify to the historical agency of women as well as men. In the process of ameliorating the burden of inequality that women bore, they improved life for everyone.

PART I

GENDER, POWER, AND LABOR

I

Putting the Barn
Before the House

"The Revolt of 'Mother,'" a short story by Mary Wilkins [Freeman] published in 1890,[1] defined the predicament of farm women in the minds of contemporary urban Americans.[2] Set in the author's native New England, the tale depicts a long-suffering wife's rebellion against her husband's decision to build another barn before building the new house he had promised her at their marriage forty years before. After Adoniram Penn steadfastly refused to discuss the matter with his wife, Sarah, she sat him down and declared:

> "Now, father, look here"—Sarah Penn had not sat down; she stood before her husband in the humble fashion of a Scripture woman—"I'm goin' to talk real plain to you; I have never sence I married you, but I'm goin' to now. I ain't never complained, an' I ain't goin' to complain now, but I'm goin' to talk plain. You see this room here, father; you look at it well. You see there ain't no carpet on the floor, an' you see the paper is all dirty, an' droppin' off the walls. We ain't had no new paper on it for ten year, an' then I put it on myself, an' it didn't cost but ninepence a roll. You see this room, father; it's all the one I've had to work in an' eat in an' sit in sence we was married. There ain't another woman in the whole town whose husband ain't got half the means you have but what's got better."…
>
> She stepped to another door and opened it. It led into the small, ill-lighted pantry. "Here," she said, "is all the buttery I've got—every place I've got for my dishes, to set away my victuals in, an' to keep my milk-pans in. Father, I've been takin' care of the milk of six cows in this place, an' now you're goin' to build a new barn, an' keep more cows, an' give me more to do in it."

"Now, father," said she, "I want to know if you think you're doin' right an' accordin' to what you profess. Here, when we was married, forty year ago, you promised me faithful that we should have a new house built in that lot over in the field before the year was out. You said you had money enough, an' you wouldn't ask me to live in no such place as this. It is forty year now, an' you've been makin' more money, an' I've been savin' of it for you ever since, an' you ain't built no house yet. You've built sheds an' cow-houses an' one new barn, an' now you're goin' to build another. Father, I want to know if you think it's right. You're lodgin' your dumb beasts better than your own flesh an' blood. I want to know if you think it's right."

Again he refused to answer.

When the new barn was completed but before the cows arrived, Sarah sent her husband off to her brother's to buy a horse. In his absence, she and the children moved their household goods into the new barn and took up residence. When the new cows came, she directed the hired hands to put three of them into the old barn and one into the empty house.

Their young hired man spread the story "all over the little village. Men assembled in the store and talked it over, women with shawls over their heads scuttled into each other's houses before their work was done. Any deviation from the ordinary course of life in their quiet town was enough to stop all progress it in. Everybody paused to look at the staid, independent figure on the side track. There was a difference of opinion with regard to her. Some held her to be insane; some, of a lawless and rebellious spirit." The minister came to remonstrate with her, but she told him:

"I've thought it all over an' over, an' I believe I'm doin' what's right. I've made it the subject of prayer, an' it's betwixt me an' the Lord an' Adoniram. There ain't no call for nobody else to worry about it....I don't doubt you mean well,...but there are things people hadn't ought to interfere with. I've been a member of the church for over forty year. I've got my own mind an' my own feet, an' I'm goin' to think my own thoughts an' go my own ways, an' nobody but the Lord is goin' to dictate to me unless I've a mind to have him."

When Adoniram returned, he was surprised to find his family living in the new barn instead of the house. Pale and frightened, he gasped: "What on airth does this mean, mother?" " 'Now, father,' said she, 'you needn't be scared. I ain't crazy. There ain't nothin' to be upset over. But we've come here to live, an' we're goin' to stay here. We've got jest as good a right here as a new horse an' cows. The house wa'n't fit to us to live in any longer, an' I made up my mind I wa'n't goin' to stay there. I've done my duty by you forty year, an' I'm goin' to do it now; but I'm goin' to live here.' " After supper, she found him sitting outside, weeping silently. "Adoniram was like a fortress whose walls had no active resistance, and went down the instant the right besieging tools were used. 'Why, mother,' he said, hoarsely, 'I hadn't no idee you was so set on't as all this comes to.'"

The urban dwellers who read this story in *Harper's Monthly* or in *The New England Nun,* which was published the next year, saw this tale as an indication of the backwardness of rural society. Like many prosperous Americans at the turn of the twentieth century, they regarded the material progress brought by industrialization and urbanization as evidence of the superiority of capitalism and the white, middle-class gender order, which not only confined women's work to the home but embellished domesticity with a wide array of consumer goods. From this perspective, rural women's lack of household conveniences and their burdensome physical labor were signs of their deprivation and degradation.

Readers attuned to the burgeoning women's rights movement read the story somewhat differently. For them, the key problem was not the material conditions of rural women's lives but the fact that men disregarded their welfare and concerns. Despite her circumstances, Sarah Penn was a model of feminine domesticity:

> She was a masterful keeper of her box of a house. Her one living-room never seemed to have in it any of the dust which the friction of life with inanimate matter produces. She swept, and there seemed to be no dirt to go before the broom; she cleaned, and one could see no difference. She was like an artist so perfect that he has apparently no art. To-day she got out a mixing bowl and a board, and rolled some pies, and there was no more flour upon her than upon her daughter who was doing finer work.

The crux of her difficulty was that her husband failed to consult her and refused to listen to her. Early in the story, she told her daughter that "men-folks" paid no attention to the opinions of "women-folks": "we know only what men-folks think we do, so far as any use of it goes, an' how we'd ought to reckon men-folks in with Providence, an' not complain of what they do any more than we do of the weather." Unable to make her voice heard, Sarah took direct action.

Ever since this powerful story's publication, the cruel irony of housing the milk cows better than the family has impressed urban readers with the patriarchal power of male farmers. It did not, however, ring true for farming folks themselves. Remarkably, Mary Wilkins Freeman herself critiqued her story in 1917. In an interview with the *Saturday Evening Post,*[3] she disparaged its lack of realism: "In the first place all fiction ought to be true and 'The Revolt of "Mother"' is not true....New England women of that period coincided with their husbands in thinking that the sources of wealth should be better housed than the consumers. That Mother would never have dreamed of putting herself ahead of Jersey cows which meant good money." Women in the Nanticoke Valley shared this sentiment. They were well aware that cows were productive property and that constructing a barn large enough to accommodate more cows was a wise investment. They saw the barn, not just the buttery, as their own workplace, for many milked twice a day alongside their husbands and children. At best, a woman working in the kitchen could stretch the income yielded by the dairy herd.

Freeman also recognized that the story's portrait of gender relations, as shown in the interaction between husband and wife, was self-contradictory. Rural women were not as passive as Sarah Penn pretended to be before her singular act of rebellion. Indeed, "If Mother had not been Mother, Father would never have been able to erect that barn." Of necessity, farm women were full partners with their husbands. "Women capable of moving into that barn would have had the cottage roof raised to insure good bedrooms. There would have been wide piazzas added to the house, and Father would simply not have dared to mention that great barn to Mother. Father would have adored Mother, but have held her in wholesome respect. She would have fixed his black tie on straightly of a Sunday morning and brushed his coat and fed him well, but she would have held the household reins...and with good reason."

This critique has much in common with recent analyses of this story and Wilkins Freeman's other domestic tales, which focus on men's failures as patriarchs and breadwinners rather than on their excessive exercise of authority.[4] But the tendency to see power within marriage as a zero-sum game was not shared by rural women, who regarded cooperation as necessary in decision making as well as labor. When they revolted, they rebelled not against their men-folks but with their husbands and neighbors against the milk processors who paid them less than the full value of what they produced.

"He Said If We Built the Barn, the Cows Would Build the House"

The narrative of an immigrant woman who lived on a hill farm in the Nanticoke Valley articulates a very different perspective from that of "Mother" in the Wilkins Freeman story. Josie Sulich Kuzma dedicated her life to raising her family and holding on to their land. Although she had a job in the shoe factory before her marriage and did domestic work in the village after she was widowed, keeping the land in the family was her proudest achievement. The daughter of Ukrainian immigrants, Josie spent her first seven years on a homestead in Canada and the rest of her short childhood on a hill farm in the town of Berkshire, just west of the Nanticoke Valley. She and her husband Alex, whom she married at the age of nineteen, bought the farm on King Hill in the town of Maine in 1932. He died twelve years later of pneumonia he contracted while building a dairy barn in the dead of winter. "When we come here it was an old, tumble-down house, and my husband was gonna build a new one. But he says he'd better build a barn first....He said if we built the barn, the cows would build the house." But "he took sick; he had pneumonia three times, and the third time he didn't make it." Josie raised their seven children in the two-room shack and managed to pay the mortgage and the taxes on the land. Eventually her youngest son returned from military service to take up farming, and after the old house burned he built her a snug cottage with a lovely view of the valley.

Josie told her life story episodically, circling around central themes—her flight from her family into a city job and then marriage, her constant struggles with poverty and with the authoritarianism of men, and her widowhood—and never finding much resolution to the issues that troubled her. Still, hers was not a victim's tale; she presented herself as coping resourcefully with

Figure 2. "Mrs. Lizzie Stewart, Brooklea Farm, Kanona, New York," September 1945. *(Photograph by Charlotte Brooks, Standard Oil (New Jersey) collection, University of Louisville Photographic Archives. [SONJ 30319])*

conditions that were entirely beyond her control. The ambivalence that pervades Josie's fatalistic account is characteristic of the recollections of women whose childhoods were marked by extreme material and/or emotional deprivation. Children who suffered from hunger and cold, parental neglect or abuse seemed to hold no one responsible for their distress. Acts that strike the listener as cruel are described in neutral terms; even in retrospect, these women did not blame their parents for their suffering.[5] Women who endured conditions of deprivation and subordination as adults recalled similar situations in their childhood. What is remarkable about Josie's life story is the resilience she exhibited as she coped with situations she felt unable to change. She had always thought for herself, but seldom felt free to act on her ideas or even to express her opinions. Widowed at forty, she was suddenly required to rely only on herself. Determined neither to farm out her young children nor to sell the farm in which she and her husband had invested so much labor, she did whatever was necessary to keep her family together on the land.

"I Come from a Very Poor Family"

While farming forms the central theme in Josie's tale, family is a counterpoint, weaving through these episodes for better and, more often than not, for worse. Josie began her life story with vivid recollections of her early childhood in rural Canada. Her parents, who were both from peasant families, had married in the Ukraine. Her father and a male friend "left soon after they was married. He said if he likes it here, he was gonna send for my mother. And then he did." Her mother traveled to the United States with her husband's friend's wife. Josie, their first child, was born in 1907 in Lehigh County, Pennsylvania. Soon the family homesteaded in Manitoba. While her father was away during the week working as a farmhand for a dollar a day, her mother tilled their land. "My mother had just one cow and she was making hay by cutting it by hand with a sickle and carrying it to the little shanty that they built for that one cow." In Europe, she had done wage labor on farms from the age of fourteen, "digging potatoes or cutting wheat. They didn't have machinery; they cut it by hand, with a round sickle." So she was accustomed to toil. On the Canadian prairie, "we had pigs, chickens, ducks, and one cow for butter and milk. If we wanted flour and sugar, she...[would] take the butter and eggs to the store and trade it for flour." "For meat, she used to kill rabbits—stand in the doorway and rabbits, nice big fat rabbits, she would just shoot 'em." Her mother also had a vegetable garden. "She used to sow barley and wheat, by hand. We had oxen still. She had a little plow" that she walked behind. "That's what she fed the chickens. And she would sow peas in the middle of that field, so us kids wouldn't get it too soon. But we'd find it."

When her parents bought a run-down farm in Berkshire in 1916, their economic position did not improve much. Her father worked away from home, first in a factory in New Jersey and then at the Endicott Johnson Company. "By that time, he got a Model T Ford and then he was away for a week; he only come home once a week." Her mother continued to do all the farming. "She had two mules and she used to raise crops, for our own use" and for sale. "We did have some cows. She sent one can of milk to the creamery; there was farmers used to pick it up." Her mother also raised vegetables, such as cabbage, carrots, and potatoes, which her father took to Endicott to sell at the farmers' market. But she was not able to support their six children and pay the mortgage through her own labor. Josie's father seldom brought any money home from his job. He spent it on "drinking and gambling. I never seen him give her any money." He seldom worked on the farm, either. "No, he wouldn't farm it. Oh, he'd come down once in a while and help out, but not very often." Driven to desperation, Josie's mother tried working in the factory, "got rid of her little cows that she had, went to town, but it didn't work." "She couldn't read and write, so she couldn't work in the factory....So we had to go back and get a few cows and start farming. Not many cows like they do now, just a very few cows." Farming was "all she knew: garden, cows. That's the only thing. A living, what little she got. That's all."

In retrospect, Josie's main grievance was not her family's poverty, which she took for granted, but her lack of opportunity to go to school. The language barrier was a serious problem. In Canada instruction was in French; the "Slavish" people built their own schoolhouse just before the Sulich family left. In Berkshire Josie was enrolled in the district school, but she did not speak English. "Started school, didn't know nothing. It was terrible. Here I was, a grown-up girl, and little kids could talk, but I didn't even know how to talk." Only one girl made any effort to help her. "It was awfully hard. It wouldn't have been so hard if my mother had believed in me being in school, but my mother used to say, 'you're not gonna live from that book, you've gotta work for your living.' And she wouldn't let us be in the book. No, if we wanted to read in the kitchen she'd take it away or she'd chase us away." Josie vividly remembered when her father gave in to her entreaties and bought her "the first-grade book." "I knew she didn't like it. But, oh, I was...I grabbed that book, and I went to the back of the house. I knew a few words, and I knew the sound of the letters. So I tried to read. I was so happy to get that book; I'll never forget it."

Her mother did not allow Josie to attend school regularly. "No, no. When we was planting potatoes I was home for a week or two. Or if there was any thrashing or something to do, I'd stay home. She didn't pay attention to how long I stay home." "My mother'd say in the morning, 'I've gotta go to Berkshire and get some feed. You stay home and watch the younger kids.' Well, that was it!" "I was the oldest, so I had to take over more," she explained. Josie never protested: "Oh, no; they'd tell you and you'd do it, and that's it. That's the way it was." When the truant officer came, Josie's mother made her hide in the back room. "And when I come back to school, there I was, a week, two weeks behind. How you're gonna catch up with the other kids? When you're way behind, you don't know."

Josie left the farm at fifteen. "I wanted to get away from home! How many times I cried to my mother, I said I wanted to leave, I wanted to leave. She'd be washing clothes by hand, and I would be nagging her I wanted to go to work, and she'd take a wet towel and slap me around the face. She didn't even want to listen to me. 'You're staying here, you're not going to work.'" "After I was fifteen, I wasn't going to school any more." Her mother had just had another baby. "I stayed home that summer, and in August, before school started, I told him [her father] I wanted to go to work. So when he went back to Endicott, he took me." It was not legal for a fifteen-year-old to leave school and work full time. But Endicott Johnson "didn't ask for a birth certificate nor nothing, just hired me." Until her sixteenth birthday, Josie worried that she'd be caught and sent back. But she was proud of being able to do the job. "My first pay was $3.75; I'll never forget that....I looked at it and I was so happy. My own money! That was something."

Her earnings were not all hers to spend, however. In addition to buying her own food and clothing, Josie paid the $5.00 per month rent for the room she shared with her father. He worked the night shift and she worked days, so they took turns sleeping in the single bed and met only occasionally for supper. She also "helped out at home. If they had any bills, they was way behind, I gave 'em money to pay for it." The second time she discussed this matter with me, her resentment was more explicit: "It was hard....I'd save a little money, they knew about it, they'd come to me, oh, they've got a bill to pay, they'd come to me and they want it. So I pulled it out and give it to them. I didn't get nothing for it. Every time I went [home], I bought her groceries and shoes for my brothers and sisters. Anything like that they'd need, I'd buy. So I didn't have any money. That's why I wanted to get away!"

Josie's narratives of her childhood left me with the impression that her parents' marriage had ultimately broken down. They lived apart, her mother remaining on the farm while her father worked in the city. Josie's mother wanted him to get a job closer to home, but "she never persuade him nothing. He was a heavy drinker. She couldn't make him do anything....In that case, you know how they are." Josie sympathized with her father. "Yeah, I seen him when he was drinking. They didn't get along when he was drinking. But I was closer to him than I was to my mother. When he was drinking, she'd get after him and nag at him, and he didn't like it. And I didn't like it." Josie was critical of her father's failure to provide for the family. But he "tried to be good with me"; at least he let her do what she wanted. Denying what I saw as her parents' de facto separation, Josie took solace in her belief that her parents never became completely estranged. Despite their chronic conflict, her father never beat her mother. "No matter how they quarreled, how they—they didn't really fight, but quarreled a lot—no matter what they said, they was together. She stayed right with him. Well, in those days they didn't believe in divorce, not like now. Oh, if she heard anybody say separation or divorce, that word was terrible, just awful. If you heard of somebody like that, well, they wouldn't even talk about it. And now it's nothing." To Josie's parents, and to her, family solidarity was more important than the quality of their marital relationship. Her parents focused their energies on getting by. "They just worked, and didn't have time to think about anything. No matter how hard they had it. On the farm, making a living, no matter how hard it was, they just kept going. Not even thinking of leaving one another."

What was unthinkable to Josie's parents was not separation—they lived separately in fact, while remaining legally married—but rather the possibility of making a different life for themselves. Returning to Europe was equally impossible, no matter how ardently Josie's mother wished she could. "She didn't like it here," Josie declared. "She always was going back. But she never had money to go back. Never got any money to go back. And then she wouldn't leave my father anyway." Her mother's whole family had remained in the Ukraine, and she stayed in touch with her mother and two sisters. She used to talk "about the old country, about how they used to live. They used to make their own clothes....All they bought was flour and sugar." Josie's mother emphasized how different conditions were in North America, where they worked so hard but had so little. Josie thought that conditions were more similar: on the farm, her family "lived mostly like they did over there. Because we didn't have nothing, nothing good to live on, any better than over there."

"And I Married Very Poor Family, Even More So Than My Own"

Josie announced forthrightly that after working for four years she "just married a man to get away from the family. I wanted to get away from the family, from my mother and father as well as the farm. Get away from home....I just wanted to be on my own. Away from 'em, move away, is all. And I did."

The man she married, who was also of Ukrainian descent, farmed with his father in Elkdale, Pennsylvania, a sparsely populated place just south of Broome County. "I was a lonesome girl after I got there," Josie continued. "Was that lonesome! No car, or nothing, there. Come from the factory, from my friends in Endicott; go over there and work, and be stuck for three weeks in blizzard weather. Sure, I had a husband then, but everything is new; you're kind of afraid of everything. It was hard. No, nobody knew how I felt." "When I was lonely I used to go see the neighbors," some of whom were also Ukrainian.

Josie recounted how she met her husband:

> I only knew him two weeks before we was married, only two weeks! I knew his sister, but I didn't know her well; I'd seen her many times as I was walking

from work, and I'd just say hello, but I didn't even stop and talk to her. One night that girl's cousins or relations was right next door, and this girl Frances knew of Alex looking for a wife, so he came....Well, his mother died two years before he came to Endicott looking. He was a farmer, and he come to Endicott looking for a farm girl. Come to a city to look for a farm girl? But at that time many farm girls left farms like I did; at that time they wanted to get out of the farm and go to the city, not to *be* in the city, but just to get out of home, like I did. Frances come over one time with him and introduced me to him. I thought, what an Indian! He was all brown. He'd come from Pennsylvania, and he'd been haying it; he was all tanned and brown and I didn't think much of him. But Frances told me who he was. The neighbor had a picture of him. And this girl that I'd met on the street all the time, she told me he was her brother. He stayed there in Endicott—he had another sister there, too—and little by little, by golly, in two weeks we was married. I went to movies three times with him and then we got married.

Josie told this story with an air of astonishment, not so much at the fact that she married a man she barely knew as at the chain of contingencies that brought them together.

Her choice of a husband was at best a secondary matter: "Oh, I had different ones, but this one came along, I married him to get away from the family." Her husband was not the first man who wanted to marry her: "No, I had other ones, but I wanted to work. The other ones I didn't really care for, but I wanted to work and see how the world looks like....I was on the farm under a bushel basket and I wanted to get out of there. So I did. And I didn't want to get married. I had a fellow, he was young, he liked me very well, but he died...I knew him and I cared for him...but I wasn't ready to get married then. If he lived, if I worked, maybe two or three more years, we probably would have gotten married. Well, I worked for four years, and then I was ready to get married."

Alex Kuzma, too, was ready to get married. "He wanted somebody to keep house for his father and him, because his mother had died two years before that. His sister was keeping house, but she got married and left them. So he come to Endicott himself to find a girl to keep house for him." Josie knew what she was getting into. "He asked me if I'd go there on the farm. So I said, 'Yeah, I'll go.' They wanted a lady that knew how to milk cows, and I said, 'Yeah, I know how to milk cows, I milked cows at home.'" Josie conceived of husband and wife as co-workers, each carrying out his or her responsibilities and putting the farm first. The personal relationship between them was of secondary importance, or so her childhood had taught her. When Josie brought Alex to Berkshire to meet her mother, "My mother would say, 'Oh, you gonna be happy'—he was kinda happy, singing along—and she said, 'You're gonna be happy with him.' I'm sure I was happy, but I worked like a mule. Oh, dear."

Her marriage did not enable Josie to escape poverty or parental domination. Indeed, her early married life was blighted by unresolved conflicts between her husband and his father. Josie and Alex lived and worked on his family's farm in Pennsylvania for almost two years, but "he didn't get along with his father, so we left." The central issue between them was money. "His father never gave us any money, no, no. We worked the farm, we helped him with the cows, but he didn't give us any allowance money, not even for a pair of shoes!" "My husband had a little extra piece of land, and he could sell the produce and make a little money for our own selves. His father wanted every cent of it. He didn't pay us nothing, and he didn't want us to keep not a penny of what we sold." Alex and Josie "used to take a horse and buggy and sell cabbage and stuff to different people in town, and when we'd come back we had to hand him all the money. And his father would keep it. He was mean."

In a later interview, I asked Josie if she had ever suggested to her husband that he speak to his father about their keeping their own money and taking over the farm eventually. "No," she replied right away. "He'd tell me, 'You mind your own business; you haven't got nothing to say.' I didn't want a slap across my face, I had to keep still. I knew how to take life then, yeah. When we was leaving, they had a meeting there, and I said something about that; I said, well, if he'd do this or do that...and boy, Alex jumped up and gave me a slap and out of the house I went! That was in Pennsylvania yet, and I thought, Oh! How can I walk home? How can I walk to Endicott to get out of there? I couldn't say nothing. You know, after you know that you can't say nothing, you just keep still, and that's it. You're hurt once, and you know it, and it stays. Oh, dear."

Josie's husband did not stand up to his father until she nearly died and he refused to pay her hospital bill. "It took us two years there before Alex finally said enough, that's enough, we're leaving. That was after I lost the baby. Two hundred dollars for the hospital bill, and his father really got after us and said that I should pay it. I was supposed to pay it, but I didn't have any money." He assumed that she had money saved up because she had worked at Endicott Johnson before she married, but she had given all her earnings to her mother. She was in no better a financial position than her husband. Josie believed that she lost the baby because of the heavy demands of farm labor and lack of medical care. "I was working too hard, haying and milking cows. And they didn't believe in doctors in those days. My mother never went to the doctor; she had babies at home, just had them, and that was it. So I didn't know I was supposed to go to the doctor about my kidneys." Josie developed a kidney infection and then eclampsia when she was seven months pregnant. She was taken to the hospital only after she went into convulsions. "They had to operate, to save me, but the baby wasn't saved."

Alex was reluctant to break with his father; "it was his farm, and he hated to give it up." When he threatened to leave, "then his father says, 'Well, I'll sign the farm over to you.' And Alex says, 'Well, if we can't get along without signing it, how are we gonna get along after you sign it? Things won't be any better.' Alex stayed long enough that he seen that he couldn't, that it would never work." Josie and Alex returned to Endicott to work and save money to start all over again on a farm of their own.

That was not the end of their dealings with Alex's father, however, for eventually he came to live with them. When he was too old to carry on alone, he sold the farm and gave the money to one of his daughters in exchange for her promise to take care of him. But he and the daughter quarreled, "and he'd hire somebody to bring him here in the middle of the night, and stay here more than he did with his daughter." That was difficult for Josie. "How many times later when his father was with us, I'd say, 'I married you, not your father; I didn't marry your father!' His father was with us a lot, and he was a kinda mean guy. I used to holler at Alex, 'I didn't marry two of you; I married one. Why should I put up with your father?'" Josie felt she had to, though, "because he didn't have no place to go." She stood up for herself a little bit, "but I didn't boss him around, just put up with whatever he wanted to do." Josie and Alex received nothing in return for taking care of his father. She thought it wasn't fair that "there was nothing for us."

Josie also ended up caring for her mother in old age. Josie had declared her independence when she married, refusing to help her mother financially. "She expected that; she said, 'Oh, we're gonna have a son-in-law, he's gonna help us.' But he told them, 'Why, we've gotta live too, we've gotta have some money.'" Josie had relatively little contact with her family for the next several years. Her father "died young" of asthma exacerbated by exposure to toxic chemicals in the tannery where he worked. For a while her mother lived alone on the farm in Berkshire, and then she went to live with Josie's older brother in Endicott. But she spent the last five years of her life with Josie. When I asked Josie what it was like to have her mother live with her after all their previous conflicts, she replied: "Well, it had to be like that. It's a mother, and you had to take her! You just have to. She had trouble with…her daughter-in-law, and she wouldn't go to my other sisters or my other brothers. She wanted to come here because I was on the farm; the other ones were in the city. Well, my brother was on the farm, but she went there and she couldn't stand it; he didn't want her, and she knew it. I [was] the only one that's on the farm, and she said 'I want to be in the garden.' That's why I had to take her, yeah. You just had to do it; time come, you do anything you had to do."

Josie found her mother less difficult to deal with than before. "Oh, yeah, things kind of settled down. After I got married, I made up my mind I was my own boss, and she couldn't do anything, no more than I can with my own kids now. I can't tell them what to do. If they ask me, I tell them, but if they don't, why, I just think about it and let it go. My brother used to tell her…she used to boss him, tell him what to do and how to do it, and he used to tell her, 'Who's doing this, you or me?' I didn't talk to her like that;

I listened to him telling her, and I learned not to. Because she didn't like it, and I didn't. When I disliked what she was saying, I'd keep still, go do something else, and let it go....I just let things go." Josie tolerated her mother's presence by ignoring her. She fulfilled the obligations she felt were incumbent on her as a daughter without either giving into or rebelling against her mother's interference.

The Kuzmas bought the farm on King Hill in 1932, during the depths of the Depression. They had rented a farm just down the road for a year. "When he bought this farm here, I didn't want it; there was no house here and I thought, oh, I don't want it; I thought, well, I'll never have a house. When I come up here, there was just a little shack; one part of it was falling down. There was just two rooms....I thought, we'll never have a decent house. But I let it go at that." They never talked over decisions much. Josie was accustomed to deferring to male authority. When I asked who had the most say-so in the family, she responded, "He did. Because he was a man. It had to go to the man; the man is always the leader. That's the way I always believed." When Alex decided to buy the farm and she disagreed, Josie "didn't dare" say anything to her husband. "He'd just holler at me, and that would hurt me more than clobbering me. So I knew better. I knew when to disagree and when to keep still. Sometimes I did say [what I thought], but not very often. No, I had to take things as they come."

Most of the time, Josie added, "there weren't many decisions; just keep going, just struggle, that's all." She reiterated how hard it was to make a go of farming. "We had kids and we had to work really hard. The Depression was very hard. What little we got from the cows and the cabbage and all, it was very cheap. And no buildings, no barn, no house....It was very hard." "We had a mortgage. We didn't pay but $300 down, and then made small payments on the mortgage." Yet, no matter how poor they were, Alex never worked off the farm. "We had to stay together. He didn't work out. We just worked right together. Whatever we had, whatever we done, we was together." The labor that Josie and her husband shared made her marriage very different from her parents'; it was a partnership forged on the land.

"When we first came here we only had two [cows]; then little by little we got more, and when we had eight we was sending milk down to Maine." While they were building up their herd, they fed the surplus milk to the bull calves and pigs, sold them, and bought heifers with the income; they also raised their own heifer calves. Once they started selling milk, they took turns sending it down with the neighboring farmers. Later, "the truck used to come up here and get the milk." After a dozen years, they were milking fourteen cows and raising young stock as well. Everyone in the family helped do the milking. "Alex and me and [the two oldest daughters], the four of us was milking then. And [the younger boys] helped. In the morning we'd get up and go and milk, then get breakfast" and send the children off to school. In the late afternoon, "we'd all be doing the milking too." The cows calved between March and May, not in the fall. With cows freshening in the spring, they could graze on the ample pastures. In the winter they fed "just hay and boughten feed. We didn't have no silo then. And if we did raise oats, we used to trade them for cow feed."

The whole family worked together during haying season. They never had a tractor, but used horses to draw the dump rake. "We had to bunch it, and load and unload, by hand. I was right with it," Josie explained. The horses would stand still when Alex dropped the reins, "and he'd pitch it on from one side and I'd pitch it on from the other side. Then take it to the barn. Once in a while I'd have to climb on top of the load and stamp it down." Josie also "helped in the oat harvest. We had a binder that would cut it and bunch it up and leave a bundle here and a bundle there. We had to go and tie it and set it up till it dried....and then take it up to the barn and stack it before we'd thrash it." Their neighbors had a steam threshing rig that horses pulled from one farm to another. It took at least four men to run the rig, and "we had to do the meals."

Josie had a wood stove in the house. But housework was secondary to farm work. "Just done it when we had time. Other times it waited." Farm work came first: "when it was raining, we done the housework; when it was nice, we'd work outside." Josie fed the cows, cleaned the barn, and chopped her own wood. "I liked to work outdoors, especially by wood; I loved it. Cutting wood, getting wood in the woods, too. In the fall, we'd pile up some wood and bring it from the woods by the house. Then right from the pile we'd cut it and take it in the house. That's what we did at home, with my mother, and that's what we did here, too."

The family had a large vegetable garden and cultivated cabbage and potatoes. "We raise enough to sell, besides our own. When I had my big family, after I had five or six kids, it took us 22 bushels of potatoes and a barrel of cabbage a year to feed the family." Alex marketed the surplus in Endicott. "He used to take a model T that had no top on it, and load it on that. He had customers" along a regular route. "He'd go along the street, and they'd come out and order some, and if he was short he'd just take orders and come back and get some more." Josie admitted that cultivating an acre of potatoes and two acres of cabbages was heavy work. "But golly, when you're young, you can do it without even thinking. I liked to dig potatoes; I still do." They kept "about fifty" chickens, "just enough to have eggs to sell. We had a little grocery store in Maine, Tymeson's, we'd take the eggs down and trade 'em for flour and sugar."

Josie was just forty when her husband died. She vividly recalled the circumstances of his illness and death. "He built that barn that is standing there now. There was an old barn up above, so he took that barn apart and all that boards that is on the new barn is from the old barn." His father helped, as did Josie. "I was up there climbing on it....He'd need a piece of board, I had to climb up the ladder to hand it to him." She didn't feel she could say anything to Alex about building them a house. "Well, he hired somebody to dig a [cellar] hole for the house; our neighbor dug a hole for the place we were going to build." The hole was all they had when he died.

"He was sick, but he wouldn't stay in bed, he wouldn't go to the doctor, he wouldn't do anything....He was sick for six weeks." When he finally went to see the doctor, "the doctor said he had pneumonia and he was getting better, and he gave him some medicine to take. On the way home he threw the medicine over the bank; he didn't even bring it home. He didn't tell me what was what; that's why I wanted to go, but he wouldn't let me go with him." But "even if I did know, it wouldn't do me any good anyway....He was the kind that didn't believe in doctors' medicine." "He didn't take care, and nobody could take care of him." She still wondered whether she should have let him be buried in bib overalls as he had requested, rather than in his new suit. But she didn't spend any time imagining how things could have been different.

"So What More Do I Want?"

After telling the sad story of her husband's death, Josie declared: "I am proud of my family. It wasn't easy for me" raising seven children alone, "but I'm glad I've got 'em!" Her last child was born just after her husband died. "She came one month too soon! She was just a little girl; I didn't think she was gonna pull through. It was too hard for me. I was carrying water from the spring—I never had water in the house—and so I took two buckets like I always did, carrying water from there, and those two buckets were *so* heavy; they were never so heavy! The next day I went to the hospital." Her youngest daughter weighed just three pounds, but she survived. Josie spent a few weeks with her husband's brother and his wife to recover from losing her husband and to visit her little baby in the hospital. When her sister-in-law volunteered to keep some of the older children, Josie refused and took them all back to the farm on King Hill. "I had a home, kept them together home; I didn't separate them noplace."

The oldest girl was sixteen, so she got a job taking care of an old lady in the village that brought in a little cash. The next daughter "was big enough to help" in the house. The three boys were between the ages of twelve and fourteen, "too young to farm it, too young to take care of cows, so I had to get rid of them. And my husband told me, he says, 'Don't kill the boys farming it. Get rid of the cows; don't make them work.' He realized, what he went through and I went through, it wasn't gonna be easy for me, but he told me not to make them farm it." Josie was able to fulfill his wishes, not forcing her sons to work too hard in their youth, and her own heartfelt desires, keeping her children in school rather than making them stay home to help her. Still, she was proudest of holding on to the land. "I didn't wanta sell it. I got money to pay for taxes, kept it up" until her youngest son came home from the military. "He was away four years—he said he wanted bib overalls. I thought, what does he want bib overalls for, being in the Air Force? Well, I got him two bib overalls....He wanted to farm it. So I hung onto the land....And boy, I'm glad I did."

They operated the dairy for a year after Alex died. Josie's son recalled that her brother came to help, but Josie emphasized her own labor and that of her children. "When I had cows, they had chores." Both her daughters and sons milked. Josie approved of children having responsibilities as long as they were not overworked. "They've got to learn there's things to do." Although it takes a lot of time to teach children how to do things, "still they've go to learn. We used to carry water from way above those trees, from the spring. When the pail was empty, they'd see that pail no water, they'd take the pail and go and get some water."

Combining farming with childcare was difficult. "Well, I used to leave the baby in the house, go down to the barn and milk two or three cows and then go up to the house, sometimes the baby would be sleeping. That was if it was cold. If it was warm, I had the baby in the basket down at the barn, or in a cardboard box, right down at the barn. When I had my [youngest daughter], sometimes I'd come into the house and find her crying, oh, crying terrible." Josie also took the babies with her into the field: "Oh, yes, to the hay field. One time we was haying it down on this field, way down through the little woods there, my husband and I was pitching on hay. My [oldest daughter] must have been about a year and a half old or about two, and she was getting sleepy, oh, she was sleepy. Okay, she was sleepy; I carried her all the way up and put her to bed, and she 'ha ha! I'm not sleepy.' I had to carry her right back down again. Oh, dear...It was a lot of fun in those times; she was a little rascal." Generally the children stayed with Josie. Unless they were sleeping, she wouldn't leave them in the house when she went outside to work; "not too long. If I left 'em in the house, I'd check on 'em real quick. I'd go back and forth. I didn't trust the other kids." Josie did not expect her oldest daughter to mind her younger siblings. "They'd help out with one another, but I wouldn't depend on them." Josie explained that "I used to take care of my brother, but I didn't want my older ones to do that, no. That was my part."

"When I sold the cattle I finished paying" the mortgage. "I had to finish paying for it because I didn't want to lose" the land. "And to pay the taxes I had to rent the land to my neighbor; he...paid me enough for taxes. Until my boy came home to take over." Josie kept two cows for milk. "And I used to raise little calves. The butcher used to come up here and get a calf that I had. I was on the welfare for eight years. That wasn't hardly enough, so I used to raise little calves. He'd come up and get my fat calf and bring me a new one, and I'd keep it another month and change off like that. From these two cows, I'd give them plenty of milk, and they'd get fat and he'd pay me extra money and that's what I had, with my welfare money. That's how I kept going. Oh, I couldn't go out to work, not with little kids. I didn't go to work until my youngest daughter was in school all day. Then I started to do housework, and little by little I got more and more, and kept it up ever since. But I didn't want to leave her home, with whom? Alone, here? I had to be with her until she was in school. Then I wanted more work....I wrote and told the welfare people I had just the one child and I thought I could make a go with her. So they stopped sending money."

Josie cleaned houses for many of the women who belonged to the Federated Church in Maine village. "I kinda liked it, because I didn't have nothing to clean with" at home. Her own house had "old beat-up floors" that "you had to take a scrub brush and soap." "I had to work the old-fashioned way. They had electric." "It was much easier, and I thought it was really fun." She exclaimed, "How wonderful it was to use a vacuum cleaner" when she went to work for people who had hardwood and linoleum floors. "When I first went to work, I told my kids, I've never had it so good, to work with." When she went to a new house, "I'd tell them, 'I don't have anything to work with, so you just tell me what you want done'....Then I done it the way they wanted it. Even today, if I go to a new place, I tell them, 'Everybody wants their work done different'; they have to tell me how they want it done."

The children accepted their poverty, just as Josie did. "That's the way my kids were, growing up without him. They didn't ask for money; they knew I didn't have it....When you go on the welfare, you can't spend it like you want to; you just have so much for soaps and so much for clothes and so much to eat, and that was it. And if you spend it all before the month was gone, well that was it; you'd just go without. That's the way it was when we was living together." "It's a struggle in life, and you just did it." Her children were friendly with the others on their hill and visited back and forth with their schoolmates. The family's poverty made some difference, though. Josie's youngest daughter "had one friend, she took her into that little old house there. The girl come—she had wanted to stay overnight—and she said, '*This* is a *house?* This is where you *live?*' Her daughter replied, 'What did you expect—that we was gonna build a castle for you, because you was gonna come?' Oh, dear," Josie chuckled.

Josie never seriously considered remarrying. One man from Endicott took an interest in her, but "he had too many ladies that he see, so I discontinued, didn't want him anymore. He wasn't looking for a wife, and he didn't care for the kids. I had kids, and the youngest was quite little then yet, too. I thought, no. That's all he wanted, was to marry a widow, but he didn't want to care for somebody else's kids." Another time, "two men came from Endicott looking for a wife. They came up and looked the barn over; they were looking the property over to see what it was worth. Then they come up to the house. I was feeding nine children: my grandchildren, my sister's children, and mine. There was nine of 'em, eating their lunch. They came in the door and looked around; they didn't say nothing. They knew I was a widow.

But a widow with nine children? When they was leaving they said, 'Well, if you decide to get married, call us and let us know.' I didn't know their names or nothing. A widow, she must have gold around here. And she did, nine children around the table! No," Josie continued, "I knew it wouldn't work to have a stepfather. You never know who you gonna get. I've seen these cases before, and it never works." "It's no fun to be forty years old and be a widow, but when you're busy with that many kids, you just don't have no time for another man. The kids would be neglected. I think kids come first. I had a man, I had him for eighteen years, and I had kids. So what more do I want?"

The most poignant story Josie told concerned attending the Federated Church in Maine village. She had been raised Orthodox, but there were no Orthodox churches in the rural community. The only church where Josie could send her children for a Christian education was Protestant. Eventually she persuaded her husband to accompany her and their daughters. But they needed respectable clothing in order to attend. The people who went to the church down in the village "dressed up more. We didn't have any clothes, didn't have any money to buy....And they asked me, why didn't we go to church. Well, we didn't have any clothes to wear." When I asked Josie if she told them that, she replied, "You bet I told them—didn't have any good clothes to wear. We wanted to, but we couldn't." The church people did not offer them any clothing, nor did Josie expect them to. Her husband insisted on buying a suit, but "I was supposed to get a dress first. I says no, you're gonna get yourself some clothes first and then I'll get a dress. And so we did. That's when we started to go to church, when we had some clothes to wear. I had to see that he'd get his first, and then I'll get mine. Because I didn't believe it; I thought if I get mine he won't get his. And then he wouldn't go with me. That's what I was afraid." "That was the only dress I had," Josie concluded; "one boughten dress in eighteen years."

She vividly recalls the first time she herself attended church. "When I walked down there, I didn't have a good dress to wear. I had a black dress, and he [Alex] made fun of me, he said it would look like you were going to a funeral. Now, that made me feel bad, but I went....We went down there; that was all right. But going out...I know everybody was glad to see me in church. I begun to cry." "I hadn't been going noplace," she explained. "It was strange...to *see* people there." When the service ended, her older daughters "took me to the back, because they were trying to shake hands with me, glad to see me; they took me right out the back. But the following Sunday when I went again I felt better; I didn't cry. Then I stayed and got acquainted with some people." Still, "I cried all the way home. I was just so glad to be there!"

For Josie Sulich Kuzma, the most painful aspect of life on the farm was neither the physical labor it required nor the fact that poverty compelled her to live in a shack, but rather the social isolation that arose from the constant demands of farm work and the family's lack of decent clothing. To the extent that she complained at all, she deplored not their dilapidated dwelling or the fact that she had to carry water a considerable distance but her husband's father's neglect of her physical well-being when she was pregnant and her husband's striking her when she ventured to suggest they leave his family farm. She never risked receiving another blow, but kept her thoughts to herself. Josie defined her heaviest burdens in terms of excessive masculine authority, not the exigencies that put the barn before the house. If they had been able to keep the cows, she remained convinced, the cows would have built the house.

2

Women's Place on the Land

How have rural women's connections to the land shaped their sense of self, the agency they felt able to exercise, and the trajectory of their life as they look back on it? The women whom I interviewed saw themselves as profoundly embedded in kinship networks and deeply grounded in the work they did to support and nurture others; family and work were inextricably interconnected in their narratives. Neither farmers nor caregivers—much less women who are both—suffer from the illusion that individuals are the masters of their own destiny. Yet most rural women did not think of themselves as hapless victims of fate, whether it took the shape of a domineering husband or grinding poverty.

"The Revolt of 'Mother'" turns on a dramatic moment when a long-suffering woman asserts herself against her husband's refusal to take her wishes and opinions into account and radically shifts the balance of power in her marriage. Josie Sulich Kuzma's story, in contrast, is shaped by the simultaneous pushes and pulls of her desire to live her own life and her obligations to others. The conflicts that pervaded her relationships with her parents and her husband's father were never fully resolved. Yet Josie blamed no one for her predicament; at least in retrospect, she felt that she had been aware of the risks she was taking in marrying a near-stranger in order to escape her family, so it was her responsibility to make her marital partnership work by starting over on a farm of their own after they had left his father. Although outsiders might see Josie as a downtrodden drudge, subordinated first to parents who were at best neglectful and then to a stubborn husband and continually subjected to destitution and social isolation, no one who met her could imagine her as abject or an object of pity. Her enormous energy, fierce determination, passionate love of growing things, and irrepressible sense of humor made her seem larger than her barely five-foot frame and the limitations of her straitened circumstances. Her life story is a testament to her having achieved what she thought was most important. As Mary Wilkins Freeman later admitted about the fictional "Mother," if Josie had really wanted a new house, she would have found a way to have one during the first dozen years she lived on King Hill. What she wanted, though, was a farm. The land that she and her husband bought with their hard-earned savings and paid for with their meager farm income was the foundation of her productive labor, and she was proud of keeping the children and the land together after his death. This is what I mean by *agency*—a woman's acknowledging both her constraints and choices and taking responsibility for her actions.

Women's sense of agency is articulated in the form of their oral narratives, as well as in what they say about the course of their lives.[1] The control that a woman could or could not exercise over her life is not an either/or, all-or-nothing proposition but a dynamic continuum. As the women whom I interviewed looked back, they identified the intricate interplay of constraints and opportunities that had shaped their life history. In a few instances, they were sure they had made their own choices because they had defied their parents' express wishes, yet many spoke of others who helped them find the courage to act for themselves. At the opposite end of the continuum, they pointed to a few instances—strikingly, most involving unwanted pregnancies or farm foreclosures—when they felt so victimized by circumstances beyond their control that they did not hold themselves responsible for the situation in which they found themselves. Most often, however, women weighed the options they had and those they lacked in relation to the historical changes that, in their view, had both expanded women's choices and created new conundrums.

My approach to these questions is based on theoretical and methodological considerations articulated by feminist scholars. Listening to women's voices means more than empathizing with them; it also means attending closely to their syntax and the positioning of the self in their narratives as the agent or object of action, sometimes literally as the subject or part of the predicate of their sentences. Sharing interpretive authority with the people we interview means not simply accepting their self-descriptions but asking them to probe the subtexts and contexts of their stories and, on occasion, inquiring about their silences or confronting them with the contradictions in their accounts.[2] Recognizing ordinary women's capacity to reflect on their lives entails more than acknowledging that, to paraphrase Carl Becker,[3] every woman is her own historian; it also involves taking her understanding of her place in the world seriously. The proverbial everywoman is also her own psychologist, sociologist, and political economist. She holds working conceptions of social relations that define what sorts of agency she is able to exercise, what makes her labor valuable, and what entitles her to a say in decision making. These notions, which social scientists too often disembody as "ideology" or "discourse," are rooted in the materiality of women's lives over time. And materiality is always social, embedded in patterns of everyday interaction, not a set of depersonalized, abstract, structural factors. Women are, indeed, aware of the power of forces that are not under their control, but they encounter them at particular places and times and, most important, at distinct moments in their lives.[4]

If we interpret women's sense of agency through the language they used to tell their life stories and the ways in which they positioned themselves retrospectively in relation to critical life events, then everything depends on how women encountered, experienced, remembered, and recounted what they regarded as key transitions and turning points.[5] The rural women I interviewed described times when they had been coerced, when they felt they had no choices, when they were excited by new opportunities for independence, and when they took charge of their fate—all at different moments in a single life. The turning points they emphasized were not the usual matters of education and occupation, courtship and childbirth but centered on when and how they embarked on or left farming, in which work and family were conjoined. Questions of power and agency figured most visibly in how they negotiated their relationships with parents, siblings, husbands, and sons and daughters over the sharing of labor, the allocation of resources, and participation in decision making.

But most important in shaping women's sense of agency, I found, was their relationship to the land. Strikingly, women's sense of connection to the land was affected less by whether they had legal title to the property[6] or by their emotional attachment to the landscape than by the life-historical process through which they came to, or departed from, the farms on which they lived. Using these women's own terminology for themselves, which I treat both as an ideologically laden discourse and as a framework for analyzing relations of property and kinship, I discuss inheriting daughters, women who married inheriting sons, and women who founded farms with their husbands. I compare them briefly with women from families that lost or left their land or that never got a foothold on the land at all, becoming or remaining wage laborers who rented houses in the country while commuting to the city to work. Those women who chose or felt compelled to make their lives in the city matter, too, even though they are usually left out of community histories.

From the 1920s through the 1950s, rural women were often described in the popular media, and sometimes described themselves to their non-farming relatives and later to me, as "farm wives." Most, when questioned, would say that they were simply "farmers." But they were well aware that in the early twentieth century their sex and marital status made "wife" their master status—the position that defined most other aspects of their identity. On the census, income tax returns, and land titles, as well as in the dominant popular culture that originated in urban society, they were first and foremost wives, and their access to property and income was seen as mediated by their husbands, who were assumed to be the heads of household. At worst, they were called "farmers' wives" in contrast to male landowning farmers, which they found deeply insulting as well as profoundly ignorant of how farming families actually worked.[7] Many referred to themselves in more collective terms as "farm people" to avoid the conundrum of their position, which the dominant culture seemed entirely unable to comprehend. Not only did they protest that the work they did both in and out of doors was essential to the success of their farms, but they argued that mothering was also central to farming families and worried about what seemed to them the recent devaluation of caregiving.[8] Acutely aware of the contradictions between the discourses and ideologies of the dominant urban society and the distinctive way of speaking and thinking about family farming, kinship, and political economy that prevailed and made sense in the rural community, most called themselves "farm women."

Several women I interviewed protested to me, as they had said to their mothers, sisters, and neighbors many times, that they shouldn't be called "farm wives" because they "didn't marry a farm"; they had married a man who happened, or decided, to be a farmer. But others ruefully acknowledged that, in marrying a farmer, they had knowingly or unwittingly married a farm as well because the demands of the farm operation always superseded the desires of the individuals—male as well as female—in the family. Who had the power to define what was necessary "for the good of the farm" is an important question that requires a careful examination of the interplay of discourses and material relations in farming families. How women negotiated the often contradictory pulls of farm work, household labor, mothering, and nonfarm employment was shaped by their relationship to the land.

In the nineteenth century, landowning families in the Nanticoke Valley had aimed to help each of their children get started in farming nearby so that as adults they could occupy a similar social position in the community, as well as to pass the home farm on to one or two heirs. The dominant American norms regarding inheritance treated all sons as equally entitled to real estate and daughters as entitled only to "moveables," personal property they would take with them to their marital households, which was considerably less valuable than land. Those laws and customs comported uneasily with rural residents' determination to keep all the land in the family and pass on the farm to a child who would carry on. Farmers with smaller amounts of land and little capital were never able to achieve the goal of helping all their children get established. They quickly adopted a system of impartible inheritance, with one son or, lacking sons, a daughter inheriting the land in exchange for caring for their aging parents; the rest of the children, most often the older sons and all the daughters, had to make their own way. Farmers with more land, but still not enough to divide among their children, shifted to a system that rural sociologists call "favored heir plus burdens"—one son or, lacking sons, a daughter would inherit the land and pay off the other heirs over a period of years.[9] There were significant ethnic variations. The Scots on Mount Ettrick gave daughters as well as sons equal shares of land; the Irish Catholics along New Ireland Road often lived out their days in sibling sets, with one sibling marrying and the others remaining single. Unmarried women who remained in the rural community customarily lived with their parents and then with married brothers or sisters. Most important, farmers kept the land in the family without married couples of two generations forming a single household. Parents and an inheriting son or daughter and his or her spouse and children might share a house, but most often they did so by dividing it in two, adding on a second kitchen and partitioning the living and sleeping space both upstairs and down. They cooperated closely in the work and shared the income, but most believed that day-to-day relationships worked better when each generation had its own household.

After the turn of the twentieth century, it became much more difficult for parents to pass viable enterprises on to their children. Doing so required them to expand the farm operation by investing more capital in land, livestock, buildings, and machinery, which many were unable to afford. Those families who were successful in making a seamless intergenerational transition generally spent years, or even decades, working in partnership between parents and adult children, and in many cases siblings continued to work in partnership once they had taken over. Yet the same economic conditions that made it difficult for farms to expand enabled some couples to buy land and get started in farming independently. Farms that had been developed by long-settled families who had departed for waged jobs in the nearby urban area were available for rent or purchase at surprisingly low prices. Some working-class families even traded a city house for a run-down farm. It was not easy to make a go of farming under these conditions, however, because couples who were struggling to pay a mortgage often lacked the funds to expand their operations. Whether natives or newcomers, farming families were in a similar predicament; intergenerational continuity on the land could no longer be assured.

Inheriting the Family Farm

Most daughters who inherited their parents' farm had no brothers, or at least none who remained in good standing with their parents into young adulthood, and brought a husband into the operation. This position gave them a substantial degree of authority, although little autonomy.

Figure 3. Dudley farmstead on both sides of Nanticoke Creek Road, Maine, New York, August 1915. "Notice Mrs. Dudley in from yard at corner of house on right, cabbage in garden." Photographer unknown. *(Courtesy of the Nanticoke Valley Historical Society.)*

Sarah Briggs Brookdale, like other inheriting daughters, fused a multi-generational family history with her autobiography. As she often pointed out, her ancestors were among the earliest white "settlers" of the Nanticoke Valley and the land they cleared has remained in the family for six generations. She inherited the farm because she was an only child; her mother had married the only child of the Briggs family when she was thirty-nine. Sarah, born in 1894, grew up under the watchful eye of her paternal grandfather, whose house they shared. She repeated to me stories that she had heard from him in exactly the same way he had told them, using pronouns that reflected his position in the lineage rather than her own. Sarah was also familiar with her mother's ancestry, but she seldom spoke of her maternal kin. When I asked why, she replied, "I divorced my mother's family." To explain, she repeated her mother's bitter litany of stories about how she had worked as hard and earned as much money as any of her brothers, contributing to the family budget and tending to ailing relatives as well as supporting herself as a dressmaker, but had been entirely left out when it came to inheriting the family's land. In her own life, Sarah found meaning in intergenerational continuity on the land and relied on the resources that her parents' class position allowed her. Indeed, a sense of the family's distinguished past, along with her parents' commitment to education and culture, sustained her when cash was hard to come by.

Like her mother, Sarah had an independent career and married a younger man when she was in her thirties. The idea of pursuing any profession or relationship that did not enable her to continue the Briggs farm was simply unthinkable, so central was her position as an inheriting daughter to her sense of self and her place in the world. In discussing courtship, she explained that when she was in her twenties she began to be courted by men "who were looking for a wife," but "I didn't want to be picked up on that basis." Another inheriting daughter of the same generation expressed appreciation that "I had a home; I didn't have to marry just anyone just to have a home." Sarah chose a man who came from a working-class family, had no relatives who depended on him financially, and was willing to take on the responsibility—as she saw it, the privilege—of sustaining her family farm through what turned out to be hard times. His successful career as a corporate executive never interfered with the farm; indeed, he helped sustain it.

Sarah and their children spent winters living in the city where her husband worked, but they spent every summer on the farm in the Nanticoke Valley. Caring for her aging parents, supervising the tenant family who did most of the manual labor, and managing the enterprise were year-round responsibilities that she and her husband shared. He not only became a well-respected figure in the rural locality but developed a keen interest in the history of the Briggs family and farm, compiling a genealogy of his wife's English and Scots-Irish ancestors and collecting their land deeds and wills from the 1790s on. Both were grateful that Sarah's grandfather had managed to reassemble all the land that had originally belonged to her line of Briggses and were committed to handing it on intact, refusing to sell their valley bottomland and ensuring that the "improved" road bypassed the house.

When I interviewed the Brookdales in 1981, they had passed the farm on to a daughter and son-in-law, who lived in the old tenant house across the road. The Angus steers grazing in the old pastures, Sarah explained, "made the place look agricultural" even though they had sold the dairy cows. Sarah and her husband lived in the lovely Greek Revival farmhouse her great-grandfather had built, filled with worn Second Empire furniture that had by then become antiques. Visitors who came to tea might hear about the lean years when people joked that the men from neighboring farms who came to help during the harvest could admire the blue-and-white china because there was so little food on it, but exhibiting rather than using those Chinese export dishes seemed as silly to Sarah as letting pastures that had been so laboriously cleared grow up to brush. Locating herself in a history that was embedded in the land, Sarah spoke with a kind of authority that only inheriting daughters exhibited.

Carrie Northfield, like Sarah Briggs Brookdale, interwove her own life history with the story of her family farm. But, unlike Sarah, she did not marry. "It's been kind of a unique circumstance," she explained to me. "A lot of people have said it's very unusual" for a single woman to have a family farm as well as an independent career. At the age of seventy-eight, Carrie was living with her brother and his wife on the farm in the town of Nanticoke that her parents had bought when she was ten. Her eyes sparkled with pride as she reflected others' praise of their unconventional and remarkably amicable arrangement. For more than fifty years, while working full-time as a teacher, she owned and operated the dairy farm in partnership with her elder sister, who also remained single, and her younger brother and his wife. Two of her nephews and their wives were gradually taking over the operation. Carrie emphasized the cooperation required by their extended family enterprise: "I've always said you can get along with anybody if you want to. Of course, you have to make some compromises, but you don't have to compromise so that you're always on the down end." Her brother Richard and his wife "are both very reasonable," and "Richard and I are a lot alike." Carrie had kept her independence but remained an integral part of a family farm partnership, so her life was "a little different from the usual run of the mill" for rural women of her generation.

Unlike the Briggs farm, the Northfield farm had not always been in the family. Carrie's father had inherited "nothing from his people, nothing.... I don't think his father ever owned a farm, although his grandfather did." "Dad's people were awfully poor," and as a youth he came from Pennsylvania to live with his married sister, who lived near Glen Aubrey. Her parents met when her mother came from Binghamton to visit her cousin, who was his sister's husband. They married in 1902 and "set up housekeeping" in part of a house their relatives owned nearby.

Like other inheriting daughters, Carrie recounted the history of the Northfield farm from her parents' viewpoints, interspersed with her own childhood memories. "My father worked with my uncle; first they had a creamery, then they had a big portable sawmill.... But my father's family was growing—he had five children by then—and what he made wasn't really enough to take care of the family. So he decided he'd have to get into something else, and he found a farm that he liked." Here she shifted to her mother's contribution to the enterprise. "Mother inherited some money," which "helped buy the farm." Her parents had a profitable market garden in Binghamton and sold some land about that time. Later, when both her parents had died and her brother bought out her share of their land, she inherited more. "I know once they used it to buy 20 acres to add onto the farm," Carrie recalled. It always "went into the land and machinery; in the beginning it bought all of that. But it was her money, and I think it was made very good use of."

The Northfields moved to the farm in 1915. "My mother never went to the barn, she never milked a cow, she never handled any hay, she never run any of the machinery, she never did any of that, no. Now, her mother hadn't...." But "she loved to work in the garden," as she and her mother had done before. The whole family pitched in: "We kids used to weed the onions and the carrots. I especially liked to weed the carrots.... One of the boys would sit on the horse—we had a horse-drawn cultivator—and Dad would hold the cultivator." Carrie's mother also kept chickens. She would "gather

the eggs and feed and water them. And we helped with that, too." "During
World War I, our neighbor next door…had twelve dozen eggs that she took
to the store to the Point, and she got twelve dollars for them. We thought
that was a wonderful thing.…They were a dollar a dozen; I know everybody
was dumbfounded. But of course it didn't last long."

Carrie ruefully recalled her one and only attempt at milking by hand:
"Well, the first day, right along, I tried to milk, and my sister Hilda tried.
Well, she could milk quite well; her hands were stronger or something. But
I couldn't. It took me half an hour or more to milk a cow. And my father
said, 'No more milking, you'll ruin the cow.' So I never tried again, ever."
Her father didn't try to teach her. "Because I had a brother two years younger
than I, and he was very energetic and he wanted to do things right away."
Her sister and brother did the milking and chores with their father; then
the three younger brothers joined them in the barn. Later Carrie learned
to use milking machines. "During the war years we had difficulty finding
help—that was bad, that was very bad—and I went to the barn then. We
had the milkers that you put on the cow and take off; then the milk had to
be emptied into pails and carried to the milk house. Well, I did that some,"
even when she was teaching every day.

As Carrie and her brother Richard recounted the history of the farm,
they took turns telling the story and even finished one another's sentences.
The enterprise did relatively well during the 1920s, when they were milking
20–25 cows. They sold Grade A fluid milk through the Dairymen's League,
first to Bordens and then to the Dairylea cooperative. They "made milk both
winter and summer, more than most people did," because milk prices were
higher in winter. But that meant they had to buy a lot of feed. "The only
thing they used to raise is oats, and corn for the ensilage." Carrie recalled
a time when milk brought less than $1.00 per hundredweight and, "with
the high cost of feed, it was impossible to make any money." They had a
lot of land in hay. "Hilda used to drive the horses to unload the hay in the
barn," Richard recalled. Carrie never did, since she was afraid of the large
draft animals. Later, during the wartime labor shortage, she drove tractor
and baled hay. Before they bought machinery, they "changed work" with
farmers in the vicinity at haying and harvest. The men who came to help
expected ample meals. "That was a nightmare," Carrie exclaimed. "Dad
always asked them for supper" in the evening, as well as dinner at noon. "I
used to want to even kick him! He always wanted vegetables and potatoes
and meat, beef or chicken, or we might have to go out and buy them pork
chops.…And pie—all the pie they could eat, and not just one kind of pie
either. Sometimes we'd have them for five or six days. That was awful!" Carrie was relieved when she was away teaching during that demanding time.

The house had few conveniences at first. "We didn't have any electricity
or any plumbing in the house when we came here.…Then we got water
piped in…and we got a big sink with a double drain board in the kitchen."

Carrie paid for the sink herself out of her earnings because it made the onerous task of washing up the milk utensils easier. "We used to have to bring them all to the house and wash them all in the kitchen sink, and that was terrible. We had to do it every day. I remember how my sister used to *slave* over the milking machines, because they had to be absolutely clean." Eventually they put running water, a hot water heater, and a sink in the milk house. "We got electricity the year I graduated from high school, in 1923," Carrie recalled. "Once we got the electric on, everything started to come forward," Richard continued. "Dad got an electric motor and we put a line shaft up in the barn, and put in a vacuum pump to run the milking machines." Carrie took up the theme: "The electric milking machine, that was a big help." They bought household appliances at the same time, including a powered washing machine. When I asked how the family could afford to buy these modern appliances, Richard explained that their father did not go into debt for them. "Usually he'd sell a cow…It was just like a revolving credit account. When he needed something, he'd sell one."

Carrie, Richard, and Hilda played different roles in the family economy. All the sons, except for Richard, worked off the farm as well as in the enterprise, but they did not contribute any of their wages to the family budget. Carrie, who also did both, frequently spent some of her earnings for things the farm needed. She had to attend normal school to get a permanent teaching license. "So I was saving my money, every penny I could.…But just before I was ready to go, Dad needed more cows; we just needed the cows so bad. And a neighbor wanted to sell his herd. So I took my money that I'd saved for school and gave it for the cows. It didn't buy them all; I imagine he still owed on them. But Dad paid me back every week.…You'd get your money saved up for something, and you might have to spend it on something else! I did that a lot of times. I was never able to accumulate" any money.

In 1933, the death of the Northfield's eldest son from a brain tumor at the age of twenty-eight devastated the family. Carrie said, "It was a terrible shock. Mother just kind of went all to pieces" and never really recovered from the loss. "She always felt pretty blue over it," Richard commented. "And she couldn't work," Carrie continued. "She tried to keep on running the house," but "she was sort of disorganized." Fortunately, "Hilda was always here, and Hilda was the mainstay. If she hadn't been here, I don't know what would have happened."

By then, Carrie had been teaching in another part of New York for three years. She became a teacher by default. "I remember my father saying, what did I expect to do? Well, of course I was going to be a teacher. I didn't want to be a teacher, but you were only a teacher or a nurse back in those days, one or the other. Well, I had no intention of being a nurse. I knew I wanted to work. And once I got started, I liked it well enough.… " When Carrie finally graduated from normal school in 1930, she took the first job she was offered. After her older brother died, "I had to come home," so she got a job in Binghamton. Because teachers were required to live in the city, she stayed with her mother's brother and his family during the week and drove to the farm on weekends. "And then I got to going back and forth every day, because I was needed at home."

Eventually Carrie tired of teaching and considered going into another profession, but too much training would have been required. In retrospect, she thought that a more demanding job would not have meshed well with her family and farm responsibilities. "You see, there were so many things that interfered with my career—my living at home here, having parents that needed some care, and having the war interfere," when her labor was urgently needed in the barn and the fields. Carrie had no regrets, however. The most fulfilling thing was raising her nephews and nieces; she regarded them "just as if they were my own children."

When I asked how it happened that neither Carrie nor her sister Hilda had married, she replied that her parents "just let things happen as they would." As she grew older, "Mother didn't give me any pressure" about remaining single. Who would succeed their parents as the operators of the farm was an open question after the oldest son died; it "just turned out that" Richard and his sisters did. Carrie characterized her father as "an easy-going man; too easy-going, actually. He didn't try controlling the boys much....Perhaps he could have encouraged them" to stay on the farm. Instead, they "took off on their own" and went into nonfarm occupations. That left Richard, the youngest. "I don't know whether he really wanted to farm or not," Carrie reflected, "but he was the only one here, and he had to help Dad."

At several key turning points, Carrie compared her life with her sister's. Hilda left school after the eighth grade. It seemed clear to Carrie that her mother was determined to have one daughter stay home to help after the birth of her sixth child in 1920. "Mother always wanted me around, and she put pressure on me then to stay home from school." But Carrie resisted, and that fall she returned to high school, boarding with her mother's parents in town during the week. Hilda, who was sixteen at the time, was also home all summer. She "was just crazy over that baby" and "that fall she didn't go back to school. She stayed here and helped Mother. Of course that was a great help to Mother, but it was not very good for Hilda." Later, Carrie returned to this painful matter: "I think, in the beginning, Hilda didn't want to" stay at home as their mother wished. "I think that she kind of, well, I *know* she did, she pressured Hilda to stay home, and she didn't really want to. I always felt bad about that. But there was nothing I could do."

Carrie felt better about her sister's life in retrospect.[10] "It depends upon how you look at it," she elaborated; "it was a lot better than marrying some person or other, just somebody around." "I think she enjoyed her life. She had a very busy, fruitful life," and she found living and working on the farm rewarding. Hilda was active in farm women's organizations and became especially friendly with a home economics extension agent from Cornell. When I commented that, without a career of her own, Hilda was less independent than Carrie, she replied laughingly, "She was less independent, but it got so that in later years she ran the business and wrote the checks." Hilda took over that role after their elder brother's death, and for decades she served as the farm's business manager. "So she kind of ruled over us all," Carrie said with a smile.

When their father was getting too old to work and their mother had died, all three siblings were living on the farm, and Richard and his wife were raising their family there. "Dad deeded the place over to Richard, Hilda, and me, before he died. It was heavily mortgaged" because "you had to either get out or get bigger." Over time, the three heirs paid off their two nonfarming siblings. "My name is still on" the deed to the land, Carrie concluded proudly.

The interplay between choice and constraint in Carrie Northfield's narrative is especially striking. Carrie looked back at her life with a clear awareness of what had and had not been under her control. She had resisted some of the pressures her parents had brought, refusing to quit school to care for her baby brother, but she recognized that in other instances none of them had any real choices. Most often, her viewpoint bespeaks a shared commitment to the family farm that meant allowing its demands to take priority over personal goals, such as when she invested her savings in the cows that "we just needed so bad." She did not even say that her father had asked her to loan him her savings; she simply knew, and did, what was necessary. In her independent career, however, she resented how few occupations were open to women at that time. She felt comfortable with the realization that her ambitions had always been limited by her responsibility for the family farm. Other important matters "just happened" or "worked out" as they did without a deliberate decision. She never discussed remaining single as a conscious choice, but she did not consider her singleness anomalous and was delighted by her close ties to the next generation. The future of the farm was uncertain after her elder brother's untimely death, but the three siblings' shared commitment to the enterprise was ultimately recompensed. Carrie Northfield's narrative demonstrates the unusual combination of autonomy and connection that having her name on the land afforded her.

Marrying an Inheriting Son

A woman who married an inheriting son accepted responsibility for sustaining her husband's family farm and transmitting it to the next generation. The situation of those whose husband worked in partnership with his father and/or brothers reveals in particularly acute form the position occupied by many women who married inheriting sons. Women who married into a landowning lineage were "outsiders," even if they had grown up nearby and their families "had known one another forever"; some felt they were never accepted in quite the same way as their children. Still, many women who married inheriting sons happily merged their life histories with the story of the farming family into which they married. As Louisa Woods Brown put it, the land "has been in this family always." The new identity she developed as a wife and mother was more securely rooted than the identity offered by her own parents, who were struggling on a rented farm. When these women told me about their lives, they started not with themselves but with their husband's ancestors who had founded the farm to which they moved at marriage. Then they looped back to tell me about their own family, childhood, and courtship, as if their lives were small tributaries that flowed into a long deep river.

Almost all these women said that the most difficult problem they faced was getting along with the relatives with whom their husband worked in partnership. Several told me that they had to be careful not to seem to assert themselves, express their opinions too vocally, or even appear to have their own ideas about the farm business because they might be blamed for whatever conflicts arose between their husband and the rest of his family. After all, parents, sons, and brothers were not supposed to argue, so they deflected the responsibility for their own disagreements onto the wives' supposed interference and undue influence.

Yet some women who married inheriting farmers were able to translate their marginality to the family enterprise into a measure of personal autonomy by establishing separate market-oriented operations. Louisa Woods Brown was careful not to "poke her nose into the business" that her husband ran with his father and brothers, but she had her own small herd of Jersey cows and made high-quality butter, as well as cultivating berries for sale. She even had a hired girl to do the kitchen and garden work as well as childcare during the summers. Several other women in this position also had small operations that were conventionally called "sidelines" because they were assumed to be less profitable than the farm's main operation; nevertheless, these were valuable sources of income that, crucially, were entirely under the women's control. Poultry flocks were especially important. Women with their own market-oriented operations, much like women who held jobs off the farm, often devoted their earnings to educating their children and ensuring that their standard of living would not suffer in comparison to city life—or, often, to making up for shortfalls during recurrent financial emergencies. Most important, they gently persuaded their husbands to pay their growing children for their labor and let them have a greater say in farm operations as they matured. These parents, unlike the previous generation, tended to set up formal business partnerships and allow their sons to buy them out at a younger age, reflecting both their own difficulties with parental control after they married and the increasing financial stresses under which farms operated. Women who married inheriting sons were especially adept at negotiating interpersonal relationships that preserved the peace in the extended family and ensured the intergenerational transfer of the enterprise.

Founding a Farm

The position of women who embarked on farming with their husbands was very different from that of those who married inheriting sons because their farm was founded on a marital partnership. Many of the couples who founded a farm together espoused an ethic that I call mutuality and they called "togetherness," making decisions jointly and sharing labor in a flexible manner that took individual inclinations into account. These women articulated a view of family farming as an equal partnership between spouses and offered their children more autonomy than those who inherited land or married into landowning families could do. None of these couples passed on the farm operation to a child, even if they had been successful financially; the children were "not really interested," they explained, and they "would never put any pressure on them." They thought that the next generation had to make their own way, as they had done. Women who saw themselves as farm partners did, however, see the land as a secure foundation for their own family.

Many of these women contributed to the down payment for the farm. They had held jobs in the city, supported themselves, and saved money from their earnings before they married. Although some had planned to use their savings to furnish a home, those who went into farming often invested it in land instead. A few, like Carrie Northfield's mother, invested money they inherited from their parents at or after their marriage. These women were likely to own the farm jointly with their husbands; unlike those who married inheriting sons, their name was on the deed. Theirs was a partnership not only by virtue of their marital status but also in terms of property ownership.[11]

Elizabeth Wheaton Graves exemplifies this sort of farm partner. She and her husband, Kenneth, had few advantages; both had lost their mothers in childhood, grew up in relative poverty, and had no access to land. Elizabeth's lifelong efforts to create a "warm family feeling" form the central theme of her autobiography. She repeatedly declared, "I've always tried to foster a family feeling. Because I had so little of it growing up." To her, the farm she founded with her husband signified security for a family that extended outward from children and kinfolk to neighbors and friends.

Born in 1905, Elizabeth lived in Maine village; her father worked in a lumber mill and as a teamster. Her mother suffered from rheumatic heart failure and died when Elizabeth was ten, so she helped care for her younger brother and run the household from an early age. After graduating from high school, she held an office job in town. Kenneth, whose mother had died and whose father had remarried and moved away, spent his childhood on his grandparents' farm. But after his grandfather died, his grandmother passed the land on to his uncle, so Kenneth learned a skilled trade. Elizabeth's father's death and her marriage formed a single dramatic episode in her life story. Two days before their planned wedding, her seventy-five-year-old father died of a heart attack. Grief-stricken, she considered postponing the ceremony, but friends assured her that her father had told them on his deathbed that he wanted her to be married "just the same" so she "would have a home."

For the first seven years after their marriage, the Graveses lived in the village while Kenneth worked in the city and Elizabeth cared for their three children and earned money by taking in washing. When they got a chance to buy a broken-down farm that had been seized for back taxes, they jumped at the opportunity even though they were not sure they could make a go of it. They took a mortgage in the depths of the Depression. Elizabeth continued to do laundry for pay, developed a market garden by herself, and made butter and cheese for sale, while Kenneth kept his job until they had managed to buy a dozen cows. They fixed up the second floor of their old farmhouse and rented it out to a succession of eastern European immigrant families. Elizabeth was equally proud of her prowess as a farmer and of making a home for their three sons, a hired hand who worked, and another hired man who was too ill to do much, as well as assorted relatives and people in distress. She took care of her husband's grandmother, her own aunt and uncle, and the elderly next-door neighbors. Elizabeth believed that the farm made all the difference between earning a mere living through waged and unpaid labor, as she had expected to do, and being able to produce enough of everything to share with others. After they sold the dairy cows and Kenneth devoted his energies to market gardening, they erected a sign that, instead of carrying the name of the original owners like many historic houses had, put their own family name on the farm.

Elizabeth shared the family's favorite inside joke. "Kenneth said—we'll have our fifty-sixth wedding anniversary in August—Kenneth said, well, one reason we've stayed together all these years: where would we go? We've had to make a home for somebody else all these years!" She laughed, and continued: "Well, we really have. A lot of people have lived with us. Kenneth always said we had to get along together, because all our relatives and friends came to live with us and we had to make a home for them! It's wonderful to be needed, you know."

Displacement

Women who were displaced from the rural community are seldom represented in local histories. I interviewed women who had grown up in poverty, whose parents were not only landless but unable to earn enough money to ensure that their children were adequately fed, clothed, and sheltered.

Although most came from families that had lived in the community for generations, they had moved about from one rented farm or ramshackle farmhouse to another and never felt that they really belonged to the neighborhood. I also interviewed women who had been neglected and abused as children because one or both of their parents had problems such as alcoholism or mental illness; a few had been abandoned and were shunted around among distant kin and foster families. Most of the rural women who had grown up in circumstances of material and emotional deprivation were surprisingly resilient. Some, like Josie Sulich Kuzma, founded farms with their husbands; a few married inheriting sons. The only women who, as adults, refused even to consider taking the risks that farming entailed were those whose parents had lost their land when they were young. Their displacement had been so traumatic that, they explained, they vowed never to put themselves in the position in which their parents had found themselves, uprooting children from the only security they knew. Most of these women moved to the city as soon as they could and never returned.

Almost all of the rural women I interviewed felt fortunate that they came of age just as women were able to be more independent. Comparing their own lives to those of their mothers, aunts, and grandmothers, they agreed that they had enjoyed greater opportunities for education and employment during their youth. Attending high school in much larger numbers, taking jobs in the city while continuing to live at home or moving to urban areas with friends and relatives, and, for most, being able to choose whether or not to farm were important new freedoms. Many young women aspired to more equal partnerships with their husbands than they had observed in the older generations. Those who remained single were delighted that they were not forced to live with, depend on, and take care of relatives, as unmarried adult women had done in the past. Substantial numbers of women who grew up in the Nanticoke Valley left, pursuing quite different and, they thought, more modern lives in the city. Those who remained in or returned to the rural community appreciated having a wider range of opportunities and felt more empowered to shape their own future, whether or not they eventually farmed.

The increasing independence of women is the central theme in Amanda Thompson's life story. Born in 1902 and raised on a farm, Amanda lost both her father and grandfather by the age of fourteen. She and her mother found the country "a hard place for a woman to make a living" and moved to town. Amanda did not marry,[12] but she remained enmeshed in a strong network of relatives and friends, many of whom were also from the countryside. She was proud that she had supported herself all her life: "You wanted to be independent, you know."

Amanda's mother was a descendent of the Scottish extended family that farmed on Mount Ettrick. Amanda remembered visiting her maternal grandmother, who had raised eight children on a two-acre plot of land she had inherited there. Her father had grown up on a farm in the same open-country neighborhood. When her parents were first married, they lived on a rented farm. Then they took over her father's parents' farm on Death Valley Road. In exchange for the land, her father agreed to support his elderly parents and pay them $25 per year in cash. Her grandparents lived apart much of the time, however; her grandfather stayed on the farm while her grandmother spent most of her time in the city with a married daughter. Amanda's grandfather died in 1914 and her father died in 1916, leaving the farm to her mother. Relatives came to help for a while, but Amanda and her mother moved to town in 1918 and sold the land in 1920. Amanda worked at Endicott Johnson and, off and on, shared a house with her mother in Johnson City.

When I asked Amanda why a woman friend of hers had not married, she replied in more personal terms: "People ask *me* that, too—'why didn't you get married?'...You know, this was just during the First World War period, and I could name at one time two dozen or more women, nice young women, who never married because there was a lot of boys who didn't come home. In fact, I know some that didn't." She laughed ruefully, and then continued, "Don't you think that was just when women were beginning to get out on their own? You see, before that, it was just like my grandmother Thompson. I loved them both and they were both wonderful people, but they just weren't too compatible. That's why she didn't stay home after she had a chance to go away." Later, when I asked Amanda whether she knew what her grandparents disagreed about, she responded, "He didn't drink. My grandmother was quite a bright woman, I know she was very good in school. I don't know just why they didn't...I guess maybe they just didn't....Women married sometimes probably men that they didn't care enough about as they should to get married." Rural women in her grandmother's and mother's generations had few alternatives to marriage: "I think in those days, a woman most had to marry or she lived home and took care of the family and was just a kind of *servant!*"

When Amanda was growing up, her opportunities were limited by her family's poverty, which became acute after her grandfather and father died. "I should have gone to school more....But you see, when I got out of grade school, we didn't have buses that took you to high school. I would have had to board in town, and my mother didn't have too much money." There was no future for a woman in the country unless she married a farmer. "But that was a period, too, when almost all the young people I knew or went to school with were all going to town and getting work. We were leaving the farms." Amanda mused that "I should have married a farmer, because I always like a farm; I was kind of a farm girl; it's still in me, I guess." She wouldn't have minded the work because "I would have known what to expect." But she never considered it "because there wasn't many" young men going into farming "around that place at that time."

Even before her father's death, Amanda knew she would have to make her own way. The Thompsons farmed 65 acres of hilly land. "We never went hungry and we never were cold," she assured me. But they had little cash; her father had to work on the roads to earn enough money to pay the taxes. "We didn't have a large dairy." They milked between six and eight cows, which freshened in the spring and grazed in the pastures during the summer. "My mother used to go to the barn and milk. And after I got older, if they were out in the field working, why, I used to help out, too. I could do it. I'd milk the cows and feed 'em, feed the chickens, you always had a pig or two, and the stock." She and her mother also helped with the haying. "That was quite a big job." As a girl, she drove the horses that powered the hay fork, which lifted the hay into the barn; her mother drove the horses that pulled the hay rake across the meadow. At harvest, "We'd go out and work in the fields. I can remember before we had a reaper and binder, and when they got that we thought it was quite a machine! That was for grain. We raised mostly oats" for cattle feed and "sometimes buckwheat. We had a binder that would gather it up and put it in bunches. I'd follow my grandfather along; he'd take a wisp of it and tie it together and put it around it and set it up. I could do it too."

Amanda described her mother as "a hard-working woman." "When the thrashers came, the neighbors would come in and help; sometimes there would be ten or twelve men. She'd start out in the morning and I don't know how she did it, to get dinner for ten or twelve men. And maybe they'd be there at night. I think thrashers must have gotten tired of chicken and biscuits! I don't ever remember much of other women coming in to help." Neither her father nor their hired man helped in the house. "You know, I wonder, I don't know whether he could have boiled water or not!" She laughed and continued: "I don't remember his ever doing any work in the house. She used to come home from town with a sick headache, and I don't know how she did it, but she'd get dinner. She used to get over it the next day, but those headaches were something that she had quite often, I know."

The Thompsons had no water in the house before 1916. "The well was about 200 yards away, down a little grade, and I remember my grandfather carrying it up by pails. Poor old man; I can see him yet carrying those pails up there; he'd set 'em down and rest a little while and then come on. My mother didn't carry the water. After my grandfather died, my father used to have to carry the water, and I guess I was getting old enough so I did some too." After her father died, her mother had a well drilled near the house. They had no electricity, and there was an outhouse around back. For laundry, her mother boiled water on the stove in an outbuilding or, in winter, in the kitchen. "The stove we had in the house was an old 'Home Comfort' that had a [hot water] reservoir on the side....Every night you'd carry wood to keep the big wood box full."

Butter and eggs were the family's main source of income. "I spent hours turning the churn to help make butter. We had a barrel churn, not a dasher. In the summertime, we'd have it in the cellar and it seems as if it'd be hours turning that thing! It's hard to do." "We sold mostly in Binghamton....Eggs and butter. We had maybe 75 chickens, not a big flock. You raised grain like oats, but not much corn; you usually bought feed for the chickens." "Every Saturday, they'd go to town—sometimes my father and mother, then later on, when I was older, just my mother and I—we had customers, we used to sell them butter and eggs and, in the fall, potatoes. I often wonder, I don't know" how the family made ends meet. "Money came in kind of hard in those days."

Her mother found it impossible to continue farming after her father died. "It would have been too hard work; the work was too hard for a woman" to do alone. And it was impossible to get reliable hired help, especially when rural youths were working in the factories and young men went off to war. An uncle came to help on the farm for a year. But then they moved to Johnson City. Her mother got a job and bought a house, where she took in boarders. Two years later, they sold the farm to relatives. "She used to be, as I was, kind of homesick....We used to get the horse and go up there once in a while." But the land was not farmed after that, and eventually it was sold out of the family.

By the time Amanda was living in the city and working full-time, she was a strong supporter of women's rights. "'Cause I really could see that many times women didn't have...Well, I was single myself and never married, and a lot of times you can see that women had to take a back seat always. I worked for Endicott Johnson all my life. I could have gotten into an office, but the main reason I didn't was because you made more money in the factory....Sometimes I wished I'd have left" the factory, "but we were fortunate that during the Depression everybody had work there, when other places many times they didn't have work. So we stayed with 'em. And by the time the Depression was over and the war was over, I was almost too old to go somewhere else. And hospitalization then was free, and they did give us lots of advantages you don't have now."

Equality between women and men at work was not among them. Amanda worked as a skiver in the cutting room, beveling the edges of the leather where it overlapped. She made samples, which paid better per piece but had to be done slowly enough to be perfect. This job was highly skilled, as was the job of the male cutters who worked at the other end of the room. Men always earned more than women, even though workers of both sexes were paid by how much each person produced rather than by the hour. "I can remember once asking for a raise, and they said, 'well, if you did that, you'd be making as much as the men!' I thought then that women didn't have the chances that men did. I could see that way back years ago." Amanda always thought that women deserve equal rights "because a lot of time a woman could do as much work or more than a man could. And women are bright. They're very capable. They didn't have the opportunity back then. Don't you think they took the women for granted? Well, they didn't realize how important they were to their lives; they couldn't have done much without their women." But all that work did not bring women much recognition.

In retrospect, Amanda Thompson was relieved that, as an unmarried woman, she had not been "a sort of servant" whose labor was taken for granted by her family. She was equally grateful that she had not been forced to marry someone she did not care enough about. For her, becoming independent meant getting a job and moving to the city. Still, Amanda faced sex discrimination in the factory that she had not known in the country, where women and men worked side by side on common tasks. Her response to the discrimination she encountered, especially her scornful reaction to the unwritten rule that no woman should ever earn as much as a man, was shaped by her life on the farm, where she knew from direct experience that women and men performed work of equal value. Amanda Thompson had to leave the countryside to support herself, but her conception of the dignity of women's labor was rooted in her childhood on the farm.

Land and Agency

Most rural women, whether they inherited, married into, or founded farms, expressed a clear sense that they could negotiate with their husbands over the division of farm tasks, participate to some degree in decisions about farming as well as childrearing, and play a central role in figuring out whether or how pass the enterprise on to the next generation. The processes through which these women came to the land on which they lived—not what work they did—shaped the forms of agency they felt able to exercise and enjoy, both in their major life decisions and in their interactions with others.

This conclusion, which is illustrated by these six women's life stories, arises from the two dozen narratives that I have documented and analyzed. The significance of land lay not so much in formal ownership, which mattered most when a farmer died or a farm went bankrupt, as in the process through which a woman came to and/or departed from the land she considered home. That process was profoundly shaped by the conjunction of her life course with economic change. Seeing the family farm succumb to debt and the Depression fractured young women's lives as traumatically as seeing a parent desert the family. Being able to pursue an education and a career without leaving the land made rural women feel liberated, especially if they remained single. Marrying into a landholding lineage was profoundly different from founding a farm with a husband. In reflecting on their lives, rural women analyzed the dynamics of power in farming families and explained the grounds on which they felt authorized to contribute to the family economy in whatever way they felt best.

The conceptions of social relations that I heard in the Nanticoke Valley were not simply personal; rather, they were widely shared. These commonalities did not arise primarily from the fact that rural women led parallel lives; indeed, they occurred in spite of class and ethnic differences. They were expressed and endorsed through kin-based and neighborly interactions that, in their combination of privacy and sociability, resembled the process of interviewing that I conducted decades later. Across their differences but with a shared rural location, these women articulated a common set of notions about the material foundations of women's power and espoused a worldview that conjoined the recognition of the value of productive labor with a range of strategies intended to ensure they were neither downtrodden drudges nor marginalized housewives but respected partners in farming families.

PART II

FARMING AND WAGE-EARNING

3

"Buying a Farm on a Small Capital"

" **B** uying a Farm on a Small Capital," published in the *American Agriculturist* in 1926, tells the story of the Young family's decades-long effort to establish a farm "under difficulties." The sixty-year-old George W. Young and one of the two sons with whom he worked in partnership, thirty-two-year-old Ralph, were interviewed by Hugh L. Cosline, the coeditor of the weekly newspaper, for a series on "the successes that had been attained by farmers" in the region. George Young's example, Cosline thought, demonstrated that a young man could start from scratch and, by "hard work" and "perseverance," make a living, pay for the land, and enjoy "the respect and liking of the entire community." "Mr. Young's farm has not made him rich. The buildings are not pretentious and the passerby might under-estimate the profitableness of the business. It is in no sense a show place, but...above the average in returns." Most important, the farm was "prosperous enough so that it was attractive for his two boys. After all, what greater satisfaction could come to a man in partnership with his two sons?" The farm not only supplied the family with "the necessities" and such "luxuries" as running water in the milk house and the kitchen but also afforded them "some available time in which to work for the betterment of the community."[1] Significantly, in rural people's eyes, transmitting the farm to the next generation and being active in civic life and farmers' organizations were as important as conducting a thriving, up-to-date agricultural operation.

The history of the Young farm exemplifies what Ralph Young called "progressive ideas"; the family adopted the latest techniques and machinery and specialized in dairying.[2] Like many Nanticoke Valley farmers, the

Youngs never had the capital, income, or inclination to expand the scale of their operation beyond what the family could handle without hiring labor. Yet the development of their farm illustrates the entire sequence of changes in methods of production, processing, and marketing that occurred between 1900 and 1945. The Youngs were progressive politically as well; indeed, their activism explains why the editor of the *American Agriculturist* interviewed them. For two generations, they were leaders in the Grange, the Farm and Home Bureau, the Dairymen's League, and the Grange League Federation, organizations through which farmers exchanged vital information, set up producers' cooperatives, and advocated for agriculture. They were equally active in the Union Center Methodist Church, which served the open-country neighborhood in the southeastern corner of Maine; in the Broughamtown school district; and in town and county governments.

George W. Young had purchased 67 acres of land in 1887, but he and his wife, Alice, did not farm it until 1904, when they had three growing boys to share the labor. The eldest and youngest sons, Stacey and Ralph, jointly purchased the farm adjacent to their parents' when they reached maturity. From 1916 through 1933 the three operated as George W. Young and Sons. After George retired, the brothers continued in partnership until 1945, when Ralph and Maude's daughter Margaret bought the farm and carried it on with her husband, Larry Coles.

The version of the Youngs' story told in the *American Agriculturist* might give the impression that the farm was primarily a masculine enterprise. As Ralph Young recounted this history, however, the women in the family—his mother; his wife, Maude; his brother Stacey's wife, Ella; these women's sisters; and the many Young daughters—were hard workers who made essential contributions to the farm. They produced the family's "living," raising and preserving most of their food and making much of their clothing; and the money they saved by minimizing household expenses was invested in the enterprise. The women also provided a vital income stream by keeping large poultry flocks and selling eggs and chickens. The adult women did not routinely go to the barn or work in the fields, although they knew how to milk, do chores, and get in the hay and would help whenever their labor was needed; the men were seldom short-handed outdoors. Yet the gender division of labor did not lead to marked divisions within the family; husbands and wives were united by common interests, made decisions about the farm together, and participated jointly in agricultural and community organizations. In their shared work and family commitments, as well as in their concern with agricultural improvement and involvement in farmers' cooperatives, the Youngs exemplified the values held by most Nanticoke Valley farmers.

The Youngs' modest success was about the best any farming family could do when they started out with little capital and struggled to make a living on the marginal soils of this upland region. Indeed, they were blessed by the conjunction of their family history with U.S. history writ large; they set up a partnership of three families, doubled their landholdings, and began expanding their operation at what proved to be the most favorable moment in the early twentieth century. Prices for agricultural commodities soared during World War I. Stacey and Ralph were able to pay off their mortgage in just three years, compared to the twenty-seven it had taken their father. Then they invested their profits in buildings, machinery, and livestock. The farm was optimal in scale, and their costs were low enough to weather the uncertain 1920s and the dismal 1930s. When Ralph and Maude's four daughters were coming of age, World War II again lifted farmers' fortunes. Margaret, the eldest, was a dedicated farmer and married a high school agriculture teacher. While her husband was serving overseas in 1945, she sold the small farm she and Larry had purchased and bought out her parents. In acting autonomously to sustain continuity on the land, Margaret exemplified the sense of self and family that farming nurtured.

In listening to the Youngs' story, we hear a tale that illuminates what many rural families hoped for but only a few of the most fortunate were able to achieve. Those who started out in agriculture after World War I were caught between the downward trend in the money they received from the sale of farm products and the upward trend in farmers' operating expenses that characterized the 1920s and were in a less secure position when they encountered the nationwide depression of the 1930s. The Youngs' practice of what they and others like them called "family democracy,"[3] their service to the rural community, and their deep commitment to collective action were certainly extraordinary. But their repeated election to leadership positions in local groups and as representatives to county-level and statewide councils of major farmers' organizations shows that many others who had less time available for travel to meetings held similar views.

From a Burdensome Mortgage
to a Thriving Partnership

George W. Young, born in Newark Valley in 1866 to Alsatian immigrants, bought land from his uncle when he was twenty-one, putting just $100 down and taking a mortgage for $1,900. The next year he married Alice, who had grown up on a farm nearby. Without livestock or equipment, they had no way of farming it, much less paying the mortgage. So they moved to Binghamton, where George had a job and Alice gave birth to three sons. In 1893, they returned to Newark Valley to care for George's ailing father, who soon quit farming and sold out. Then they moved onto their own land for the first time. But, as George explained to Cosline, "prices [for farm products] kept going down further and further and we found it slow and discouraging work to make our payments."[4] So they moved off the farm in 1901, and George went to work for his brother-in-law in the city. Three years later, they returned and struggled to make a go of it. Ralph, who was ten at that time, wrote in his memoir, "Now we would be classed as poor. We probably were then but didn't realize it."

From 1904 until 1915, the Youngs worked to establish their farm and pay off the mortgage. George realized that they had too little land to operate profitably, so he bought an additional 20 acres in 1908 and 23 more in 1912. They concentrated on dairying and poultry. Milking about ten cows from early spring through late fall, they separated the milk with a hand-cranked machine, churned, and washed and packed the butter at home. George and the three boys did these tasks most of the time, but when any two of them were away, otherwise occupied, or ill, Alice milked, did barn chores, and took care of the butter. The poultry flock was mainly Alice's responsibility. Each spring, they set up incubators in the cellar of their house and purchased several hundred fertile eggs. When the chicks hatched, they put them in the brooders. In early summer, they sold off the roosters for meat and put the pullets in the chicken house outside. The flock required a good deal of tending. Eggs had to gathered at least twice a day so they would not get broken, hens had to be fed and watered daily, and the chicken house had to be cleaned periodically to keep down mites and prevent disease. Poultry was a risky business, but the operation made a substantial contribution to the farm income. The Youngs packed eggs in large crates and sold them to merchants in Union Center along with their butter.

The farm was quite diversified. The Youngs grew hay and oats to feed their cattle as well as potatoes and a garden to feed themselves. The skimmed milk that was left over from butter-making was fed to the pigs. They sold whatever surplus hay, oats, potatoes, and pork they had. They kept their labor costs down by "changing work" with their neighbors during haying and harvest. The men in seven or eight families joined together in a crew that went around to do the haying, harvesting, and threshing (locally called thrashing) on each of their farms in succession, while the women cooked enormous noontime dinners for them. No money changed hands because all the farmers who helped others received help themselves. The Young boys earned extra money however they could; in his memoir, Ralph recalled digging potatoes on the neighboring farm for $1.25 a day. When her sons were away, Alice would draw in the hay as well as milking and doing barn chores with her husband.

Ralph occasionally helped his mother when she had too much to do. On washdays, he often hung the clothes out to dry, but he did not need to carry water after they had piped water into the kitchen. One Saturday in January 1913, he wrote in his diary, "I helped Mother get ready to go to Ladies Aid dinner at N. Pitchers. She and Father went. Stacey and I dosed hens at night. I *baked a chocolate pie.*" Ralph underlined this sentence in his diary and regarded it as a singular enough achievement to list in his memorandum of the year's important events. That March, when inclement weather prevented him from cutting wood outdoors, "I helped Mother quilt from

9 a.m. till 3 p.m.... We did about the least real work of any day this winter." By Ralph's standards, quilting, however tedious and exacting, did not qualify as "real work" because it was not physically strenuous. Later, when Ralph was recovering from an illness that kept him confined to the house, "I helped Mother and we finished tying the quilt." The rare instances when Ralph helped his mother in the house did not quite balance those in which his mother helped the men in the fields. But Ralph's diary shows that he was aware of what his mother was doing every day, as well as of what his father and brothers were doing.

In winter 1913–1914, nineteen-year-old Ralph attended a short course at the State College of Agriculture in Ithaca. Although his formal education had ended after the eighth grade because the family could not spare his labor, much less afford to pay his board in Union where the high school was located, he studied on his own and passed the entrance exam, winning a Grange scholarship. The family made many changes right after he returned. That spring they shifted from making butter on the farm to taking milk to the creamery in Union Center. They reconstructed the dairy barn that summer, putting in a concrete floor, piping water from the house to the barn, and adding a holding tank for the milk. At the same time, they built a silo in a corner of the barn so they could store their cattle feed in a way that maximized its nutritional value. George relied on his son's new expertise. In his memoir, Ralph wrote that "during the time I worked with Father, from the time I was fourteen until I was twenty-one, he would consult me before making any decision of importance. I sometimes felt I had too much responsibility, but, as I look back on it, that was probably one reason why there was no generation gap" in the family.

In 1916, just a year after George paid off the mortgage he had taken out in 1887, Stacey, who had been driving the bus between Maine and Endicott, proposed to his father that they go into partnership. George thought that the family's 110-acre farm was too small to employ both Stacey and Ralph, so he recommended that his sons "buy the adjoining farm, on their own responsibility." After dickering with the landowner, the brothers clinched the deal to buy the 87-acre farm for $2,450 and got a mortgage from a bank in town. Twenty-six-year-old Stacey and his wife, Ella, moved into the farmhouse; their first child was born six months later.

In 1917, Ralph married Maude Ketchum, whom he had been courting for four years. The Youngs, the Ketchums, and the Woodwards had been friends since the Youngs moved to Maine; all three families were active in the Methodist Church in Union Center. Ketchums had been in Maine for a century, and Maude had eight siblings as well as myriad cousins nearby. In his diary, Ralph spoke of them collectively as "the Ketchum tribe" or "the Ketchum gang." The young people socialized in mixed-sex groups, both with and without their parents. By the time Ralph and Maude decided to marry, his elder brother, Warren, had married Elizabeth Woodward and Harry

Woodward had married Maude's sister Mabel. Later Maude's brother Ray married Marion Oliver and her sister Carolyn married Marion's brother Leslie. These intermarriages ensured that the close ties among the siblings and friends in the Young, Ketchum, Woodward, and Oliver families continued throughout their adult lives.

In an interview, Ralph explained that "it was quite a while—more than usual" between the time they decided to marry and when they were wed. His diary notes that "Maude and I gave each other rings for Christmas" in 1914, a year after they began courting, but they did not marry until June 20, 1917. When I asked why they waited, Ralph said simply, "You wanted to be sure of things before you got into it." His daughter Margaret elaborated; there were "financial reasons" for the delay. It was not yet clear what arrangements Ralph, his father, and his brother would make about the farm. Because Stacey's family occupied the other farmhouse, Ralph and Maude would have to live with Ralph's parents. There was some question about the suitability of that arrangement, but the two couples decided they could cooperate. "It was a peculiar situation," Ralph concluded; "we never lived in a house by ourselves during our entire life. It just seemed to be the best way."

Once the Youngs had doubled their acreage and had three families to share the work, they were able to expand their dairy operation. By summer 1917, they were milking fourteen cows. For the first few years, most of the profits from the milk and hay they sold went to pay off Stacey and Ralph's mortgage while "the chickens bought the groceries." On September 1, 1919, he wrote, "Maude and I to Union…finished paying the mortgage and had discharge made out." They began investing in new equipment the next year. The Youngs already had a truck for farm use, and they were among the first in the vicinity to purchase a tractor. Next they installed lights in their house, barn, and henhouse, which were powered by an acetylene generator. In spring 1921, they had a new well drilled; in the fall, they bought a new silo, locating it next to the barn. Then they built a milk house and purchased milking machinery, which cut the amount of time it took to do the milking in half. When George was called away the day his father died in February 1922, Ralph "did chores alone. Are milking 19 at present. Did milking in 50 minutes with 2 machines." In retrospect, Ralph said that this increase in efficiency was important primarily because it freed them to do other work.

The Youngs took pride in being "up to date" technologically. Because they had little cash to spare, however, they made the farm machinery they purchased pay for itself by hiring out to work for others. In the winter, George and his sons earned money by cutting wood with their gasoline-powered saw; in the summer, they did thrashing and silo-filling for other farmers with their gasoline-powered threshing machine and blower. The ample supply of family labor and the women's ability to fill in for the men who were away from the farm helped ensure their success. So did the caution with which they expanded. The Youngs carefully calculated the optimal scale of operations for their land and labor, and they made sure to stay within their financial means. They realized that the costs of an improvement could readily rise more rapidly than the returns, as the outright losses and mounting indebtedness of some of their overly optimistic neighbors demonstrated. They kept a close watch on the efficient utilization of all their resources and tried to ensure a well-balanced farm operation. As Ralph explained, he and his father agreed that husbanding their assets was as important to their success as continually expanding their milk production.

An Exemplary Dairy Farm

The portrait of the Young farm in 1926 exemplifies what both agricultural experts and local farm families regarded as best practices at that time. Ralph gave the coeditor of the *American Agriculturist* a tour of the farm and explained the Youngs' approach to management. When they shifted from making butter to selling milk, they converted from Jersey cows, which produce milk with a high butterfat content, to Holsteins, which produce a larger volume of milk. Their herd average was relatively high for that time. Their cows had been tested for bovine tuberculosis under a federal program to eradicate the disease, which could transmit TB to children who drank fresh milk. Cows found to be infected were destroyed and the farmers compensated. The Youngs led the effort to convince local farmers to comply with the program.

Ralph explained their farm management strategy: "We have been trying to save on the cost of production by raising more of our feed at home." The previous year they grew 800 bushels of oats and 100 bushels of barley and planted four acres in oats, buckwheat, and field peas. "We ground this at home and bought cottonseed meal and gluten to mix with it....We figured that this made a mixture with slightly over 20% of protein." High-protein feed helped increase milk production, while growing as much feed as they could and grinding it together with purchased supplements controlled their cost. Although they adopted ensilage, they could plant corn on only eight acres; they were experimenting with hardier, more rapidly ripening varieties being developed by the Agricultural Experiment Station. For bulk feed, they relied on hay. Because they had a large amount of hilly land that was unsuitable for tillage, they had ample pasture and cut more hay than their own cattle required. "We sell about forty or fifty tons of hay each year," Ralph continued.[5] Each year they spread lime on their meadows and pastures and sowed a little alfalfa amid the timothy and clover to fix nitrogen in the soil and raise its fertility. That meant they could sell their hay at a premium price. Not only was alfalfa in the hay uncommon in this region, but not all farmers seeded their meadows; some simply converted old hayfields to permanent pasture when they were no longer productive.

"In these days of the high cost of labor," Cosline continued, "I am always interested in learning what equipment is used on the farm and in the home."[6] The Youngs' labor-saving devices enabled two people to do the milking and chores by themselves. But avoiding having to hire year-round help was more than the cost-saving measure that Cosline recommended; utilizing family labor was central to the Youngs' entire philosophy. Had George and Alice not had two sons and daughters-in-law who wanted to farm, they would probably have done what many of their neighbors did: maintained a small dairy herd and a large poultry flock, planted a big garden, and carried on as long as they could do so without going into debt or suffering actual material want. Then they would have reluctantly moved back to town and sold or rented out the farm. The labor of their three growing sons had enabled them to make a go of farming, and two adult sons' choice of farming as a vocation spurred their expansion.

The *American Agriculturist* was most interested in the Youngs' commercial dairy operation and proclaimed "Hens a Side Line." But Cosline had to admit that the "flock of 200 white leghorn hens" was "profitable." "They are fed, culled, and managed according to the recommendation of the State College of Agriculture and the average production per year is about 140 eggs per hen."[7] Indeed, their relatively large and productive poultry flock was the second most important farm operation, subsidizing the expansion of the dairy. The installation of lights dramatically increased egg production and extended the poultry cycle into the winter. Ralph marveled at the results in his diary on January 22, 1921: "Today completed 2 weeks that the lights have been on in the henhouse. On Saturday 8 Jan. we got 31 eggs. Today we got 137." When they hatched their own chicks each spring, only half survived, so in 1927 they began purchasing peeps from the McGregors. The women of the family continued to carry the major responsibility for the poultry flock, but as they enlarged the scale of this operation the men had to help, especially when they packed dozens upon dozens of eggs in crates for shipment to New York City.

The *American Agriculturist* did not mention their other cash crops because these operations were small in scale and, like the garden, tended to be taken for granted. Nearly all rural residents cultivated vegetables and sold whatever surplus they could. The Youngs produced several hundred bushels of potatoes for sale, in addition to home consumption. Planting and digging the tubers were laborious tasks, and the whole family toiled in the three-acre potato field.

Cosline treated George W. Young and Sons as if the men's partnership were synonymous with the farm, and he never even mentioned the women and children in the family. Like many experts who advocated the modernization of farming, the *American Agriculturist* had a distinct masculine bias.

Figure 4. "Farm women dressing chickens, Brooklea Farm, Kanona, New York," September 1945. Mrs. Lizzie Stewart and an older woman pick chickens in the kitchen. *(Photograph by Charlotte Brooks, Standard Oil (New Jersey) collection, University of Louisville Photographic Archives. [SONJ 30324])*

The coeditor regarded the large commercial operations conducted primarily by men as the really important work; what women did was either taken for granted or ignored, in spite of the fact that the work they did, both alongside their menfolk and independently, made a substantial contribution to the farm income. Ralph Young, like most of those who proudly called themselves "practical farmers," knew differently. In his way of thinking, women and men were full and equal partners in both family and farm, and the integration of work and family was one of the major reasons why farming remained an especially satisfying way of life at a time when city dwellers might earn higher and more stable incomes with less arduous labor. As Ralph told the story, the family was inseparable from the farm. The birth

and growth of their children, the illness and death of their elders, and their
close relationships with relatives and friends were as important to him as the
steady enlargement of their operations and the financial stability they maintained through uncertain and downright bad times. Like George Riley of
East Maine, whose late-nineteenth-century diary includes detailed notes on
domestic life as well as farm tasks and financial transactions, Ralph Young
recorded the many daily activities he shared with his wife and children and
the entire extended family.[8] His memoirs, too, are as much a family as a
farm history.

Ralph and Maude had four daughters between 1918 and 1928. For the
first year of their marriage, they had their own room in his parents' house,
but then they divided the house into two households, the young couple taking the sitting room side and the older couple keeping the parlor side. With
a baby on the way, they needed more space, and Ralph's parents may well
have wanted more privacy. As the family grew, they modernized the house.
In 1919, they bought a coal heater, and in 1923, they installed a furnace.
That fall and winter, they built a new kitchen onto Ralph and Maude's
side of the house and equipped it with electrical appliances. They repaired
George and Alice's side of the house the next spring. The Youngs did most
of these renovations themselves rather than paying skilled tradesmen. In
January 1924, Ralph noted in his diary that the project was "going slow as
we are working on it at odd times" when they did not have more pressing
tasks to do. Only when they got indoor plumbing did they hire an expert;
in 1923 they had two bathrooms installed, one on each side of the house. In
1924 they put in a new sink and bathtub by themselves.

Home economics experts recommended household conveniences to
lighten farm women's labor and praised men's willingness to spend money
on them, treating it as an indication that men considered women's welfare.
While emphasizing that his family improved their house right along with
the dairy barn and henhouse, Ralph objected to the assumption that women
did nothing but housework and men held the purse strings. He declared
that, while his father held women in high respect and his mother deserved
no less, they had a different idea of equality than the experts who spoke as
if men were the real farmers and women were merely homemakers. Like
many Nanticoke Valley residents, Ralph Young thought these ideas had
little bearing on how farm people actually lived and what they believed. In
his view, women and men had common rather than separate interests and
should make major decisions together. His parents, George and Alice, were
the first couple to sit together in the Union Center Methodist Church; until
then, the men sat on one side and the women on the other. Women were
active participants in farmers' movements, and the Youngs believed that
democracy required their recognition as full citizens. In 1915, when George
was master of the Union Center Grange and Alice, as lecturer, was responsible for arranging programs, they invited "Miss Warford, a suffragette,"

to speak during the New York state campaign. In 1918, at the height of the national campaign, Ralph noted that "the offices were all filled by the Ladies, who took charge of the meeting." After the "first election in which women voted," as Ralph put it triumphantly in his diary, Alice served on the Election Board along with him; in 1920, mother and son spent all night counting votes. In Ralph's way of thinking, real equality meant having a political voice. Just as women and men worked together to sustain family farms, so they worked together in the community. Utilities and labor-saving machinery freed up everyone's time for more rewarding work and for active participation in civic and farmers' organizations.

The Second and Third Generations

Between 1927 and 1934, when the sixty-eight-year-old George Young turned the management of the farm over to his forty-two- and forty-year-old sons, the Youngs continued to adopt innovations that made their operations more efficient and enabled them to do without hired labor.

The three couples maintained separate households, but cooperated in both farm work and childcare. After Alice died in 1933, the two sides of the house were merged. When George passed the farm on to his sons, he relinquished the decision making but continued to participate in farm work when he was not away on business for the county Board of Supervisors, to which he was repeatedly elected, or for the farmers' organizations in which he held county or statewide office. As Stacey and Ella's daughter and son and Ralph and Maude's four daughters grew old enough, they helped their parents both indoors and out. In late May 1931, for example, "We planted potatoes—Stacey and Howard, Margaret, Mildred, Jean and I. Didn't quite finish by noon, so Maude brought out dinner and we all had a picnic." Because the Youngs placed a high value on education, they ensured that all their children graduated from Union-Endicott High School, although none attended college. As Margaret put it, "the money simply wasn't available" during the Depression.

Fortuitously, the Youngs made an important shift in their dairy operation just before the nationwide financial crisis. In summer 1927 they built an entirely new barn. By buying new equipment and improving their buildings, they were able to operate on a somewhat larger scale. The most significant change, however, was that they began milking cows through the winter. That required substantially more feed, but the milk brought considerably higher prices because the overall supply declined. In the previous system, which was called "summer dairying," the cows calved and started giving milk ("came in" or "freshened") in the early spring; they grazed outside through the summer; and as their milk supply diminished in the fall, they were "dried off" until they calved again. "Winter dairying," as the new regimen was called, could be more profitable but only if the farm could avoid spending too much money on feed and hired labor. Ralph kept careful track of the cows' schedules in 1927–1928, when they were shifting the milking cycle. From early 1928 on, they milked about eighteen cows from January through June and averaged eight cans a day, making their herd more productive than those of most farmers in the upland portions of Broome County. Shifting to winter dairying when they did enabled the Youngs to maintain a relatively stable operation throughout the 1930s.

Like many farm families, the Youngs increased the scale of their poultry operation in the early years of the Depression. In 1932, after inspecting the McGregors' modern poultry farm, they enlarged their henhouse, purchased 700 chicks instead of 400, and kept 325 pullets through the winter. The demand for eggs, like that for milk, remained relatively strong despite the dismal state of the economy, and prices for these vital foodstuffs did not fall as much as those of other farm products.

In the late 1920s the Youngs acquired more labor-saving devices for the house just as they did for the barn. New appliances did not mean that each woman worked alone, however. After Maude bought a used Singer sewing machine, her mother and sisters regularly came over to use it. For two years after they purchased a powered washing machine, Maude's sister-in-law came over to do her own laundry. Ralph's diary also records the assistance that Maude provided to neighbors who were not kin. In 1929, "Maude started looking after Paul Stratton while Rena teaches at Union Center." The Strattons were old friends who lived along the Union Center–Newark Valley road, and Maude took care of their young son, along with her own baby daughter, while Rena helped support her family. Women often traded childcare, but in this instance Rena paid Maude. All these cooperative arrangements, whether they involved cash payment, an even exchange of goods and services, or mutual aid, were grounded in long-term, reciprocal relationships between families.

Changing work at haying and harvest; getting together to sew or can; borrowing and lending tools and supplies; cooperating in the marketing of dairy products and the purchasing of farm supplies; maintaining the roads; supporting the district school; and assisting one another in childbirth, sickness, and death—all knit neighbors together. Not only did women frequently visit one another to see what they needed and lend a hand, but men helped one another in emergencies without being asked. For example, in January 1920 Ralph noted: "On the way to Union Center with milk stopped at Gus Allen's and found him sick in bed with no chores done. Took milk, then called Dr. Came back and did his chores." The men along the Union Center–Newark Valley road held a "wood bee" for another man who was ill. Men sometimes cared for the elderly and infirm, performing such intimate masculine tasks as shaving. Men's and women's social relationships formed overlapping circles that brought them together in others' households.

Indeed, patterns of association among households promoted gender integration in families where the labor was more divided. In mutigenerational farm partnerships like the Youngs', where there were enough men to do the outdoor labor with only intermittent assistance from the women and enough dependent children and elders to require considerable labor in the house and garden, husbands and wives conducted a common social life and participated in community activities together.[9] The church and the Grange brought Maude and Ralph together more regularly than their daily round of work did.

Ralph's diary for March 21, 1935, reads: "Father, Stacey's folks, and us to Binghamton and made out writings for the sale of our places. Stacey bought our share of that farm and we bought Father's farm." The middle son, Warren, who had never been interested in farming, was compensated for his share of the family property; in effect, his brothers bought out his interest in their common inheritance. Although Stacey's and Ralph's families each held title to their own land, they continued to operate in partnership.

The Youngs made no major changes in the farm operation; they sold milk and eggs in Endicott, raised hay and cattle feed, and planted a garden for their own use. They delivered their milk every morning to a retail dairy, which pasteurized and bottled it for local distribution. They did not enlarge the dairy operation during these difficult years, although they raised their own young stock and occasionally bought a cow in mid-summer when their herd failed to meet the quota specified in their milk contract. They also had contracts to supply eggs to a number of stores in Endicott. They bought around six hundred peeps from the McGregors each spring and sold the roosters in mid-summer. Raising poultry was still a risky business; for example, in April 1934 Ralph "found the chick houses cold so stayed up and tended them, didn't go to bed."

The family continued to adopt new methods, especially after observing other farmers who had positive experiences with innovations. In 1935, they improved the water supply to the milk house and put in a concrete vat and floor, which cooled the milk cans more effectively and made the room easier to clean. They piped water to the henhouse at the same time. The next summer, they put a steel roof on the barn. Remarkably, the Youngs were willing and able to borrow money to finance improvements. On April 1, 1936, Ralph wrote, "Maude and I went to Binghamton, made application for a Federal Loan on this place." Through his work with farm organizations, George was well informed about New Deal programs to assist farmers. The family was confident that they could repay the loan, especially because the sum they borrowed was relatively small compared to the value of their enterprise.

Even more remarkably, the Youngs made extensive changes in the house during these years. In 1934, they installed a battery-powered electrical plant and wired the whole house. Ralph was active in the campaign for rural electrification, and main power lines were finally strung along the Union Center–Newark Valley road in summer 1936. They used some of the money they borrowed to remodel the kitchen, "installing new sink, cupboards, linoleum, etc."; "adding on a back room"; and purchasing an electric combination range and stove, refrigerator, and hot water heater. Ralph and Maude's house was as well-equipped as any in the Nanticoke Valley, and in the late 1930s, its comforts were comparable to those that Warren and Elizabeth enjoyed in Endicott.

Eleven years after he and Stacey assumed full control of the farm, Ralph decided to stop farming and go to work for a government agency that made operating loans to farmers. Following in his father's footsteps, Ralph became active in the Broome County branch of the National Farm Loan Association. In 1945, he was recruited to join the staff of the Production Credit Association (PCA), inspecting farms to see whether they qualified for loans. Taking the job would enable him to use his practical experience to serve other farmers in a time of crisis. The only catch was that he and Maude would have to move to Delaware County, at least temporarily. Stacey and Ella, too, wanted to stop farming. Ralph acknowledged in his memoir, "One reason why I was able to be active in so many organizations when I was farming is that Stacey did not care to become involved in community affairs. He would do the chores many times when I was away. To make up somewhat for that, I kept the books of the farm business on my own time." Stacey and Ella were tired of the unremitting round of chores that keeping dairy cattle and chickens required. All six of the two couples' children had married and left home. Only Margaret, Ralph and Maude's eldest, expressed any interest in farming.

By her own account, Margaret was more inclined to become a full-time farmer than her husband, Larry Coles, even though he had studied agriculture at Cornell. By 1945, Margaret and Larry had been married for six years and had two small children. Larry had taught vocational agriculture in high school, worked for the cooperative Grange League Federation, and done war work at the IBM plant in Endicott. The young couple had even purchased a small farm of their own. They came to visit her parents whenever they could. When I interviewed Larry and Margaret in 1982, he laughingly recalled how much he hated toiling in the fields during his two-week summer vacation. When Larry was finally drafted, Margaret and the two children moved into the other side of the Youngs' house for the duration, and she was there when her father decided to accept the job offer from the PCA. Unbeknownst to her husband, who was stationed in Japan, Margaret decided to sell their own land and buy her parents' farm. When I interviewed Margaret and Larry thirty-seven years later, when both were in their sixties and still farming the home place, they joked about her making such an important decision all by herself. Aware that she enjoyed her husband's unqualified approbation, Margaret explained that the Youngs' land was better than theirs; they had benefited from all the work that previous generations had put into the place. She concluded quietly, "I suppose in a way it was wanting to keep it in the family; there was no one else who wanted to farm."

4

The Transformation of Agriculture
and the Rural Economy

By 1926, when the *American Agriculturist* held up the Young farm as evidence that a young man without capital could become a successful farmer, editors and readers were deeply anxious about the future of agriculture in the northeastern United States. The farmer who had been profiled just before George Young "said that the young men of today are not willing to start under the handicaps which their fathers had."[1] Everyone understood that many men and women who grew up on farms were leaving rural areas rather than taking over their parents' enterprises or establishing their own. But the real problem had to do with economics, not with youths' unwillingness to work hard enough. Had Stacey and Ralph Young not been able to carry on a going operation, they would have found it much more difficult to buy land, expand the dairy, and shift to fluid milk production before the Depression. Since the turn of the twentieth century, farms in this region were being sold, rented, and even abandoned as families moved to town in search of a reasonable return for their labor. The advice the farm paper offered was suitable for those who had the resources that expansion required, but it ignored those who were struggling merely to subsist and pay the mortgage as well as those who, like the Italians who rented the farm adjacent to the Youngs, had only a temporary foothold on the land.

In the early twentieth century, the rural economy of Broome County was transformed by two interconnected socioeconomic trends: a structural shift toward more specialized, larger-scale agriculture, which the Young farm exemplifies, and an emerging pattern in which families combined farming with wage-earning. Under capitalist conditions, farmers had to expand

and intensify their operations to stay in business; as one farmer put it, "if you didn't get bigger, you'd go under." Financial requirements rose as the purchase of inputs such as feed and fertilizer, the processing and sale of products, and even the process of passing on the family enterprise became increasingly embedded in the market. Farmers' organizations enabled more families to survive these shifts than would have been able to do so without collective action, but cooperatives were unable to reverse these basic trends. At the same time, in what is commonly but erroneously called part-time farming, families sent some members to work off the farm while the rest conducted a range of small-scale subsistence and market-oriented operations on the land. Although the term *part-time farming* suggests that no one worked on the land full-time, that assumption is incorrect; in almost all these families, one or more individuals devoted their labor entirely to agriculture while others held off-farm jobs. These families allocated their labor between farming and wage-earning in flexible and varied ways, depending on their labor and financial resources and on the job and market opportunities available to them.

In American agricultural history, the continual increases in scale and specialization that characterized agribusiness and the emergence of part-time farming are usually seen as sequential. In many regions, people continued to leave the countryside for the city through the late twentieth century, and only recently have families who were unable to get by on their farm income sent members to work off the farm without leaving the land entirely.[2] In Broome County, where some farmers were able to keep pace with the changes that continued viability required but many lacked the necessary capital, the two trends were simultaneous, complementary, and mutually reinforcing. Nanticoke Valley residents whose resources were insufficient to expand the scale of their farm operations were able to take the bus driven by people such as Stacey Young from their country homes to city jobs, and urban workers were able to rent or buy vacant farms at relatively low cost. For every farming family that, like the Youngs, managed to stay abreast of agricultural change, three gave up and moved into town, one remained on the land but also relied on urban employment, and another was replaced by newcomers who sought to complement wage-earning with farming in a small way.

In this chapter, I trace the changing rural economy using information collected and analyzed by the U.S. and New York State censuses of agriculture and population for Broome County[3] and studies of Broome County farms conducted by experts in agricultural economics and farm management at the New York State Agricultural Experiment Station.[4] I explain the economic changes through which this stratification process occurred and analyze the connections between class position and the utilization of family labor. Significantly, class position corresponded with distinctly different patterns of intergenerational and gender relations in rural families.

Diagnosing Rural Depopulation and Economic Decline

The economic history of Broome County was shaped by its hill-and-valley topography. Compared to other areas of the Southern Tier (the counties of south-central New York that bordered on Pennsylvania), its soils were less suitable for arable farming and their deficiency of lime made even hayfields and pastures less lush. Its damp and chilly climate, combined with its poorly drained clay soils, put corn at risk as well. In the late nineteenth century, after the hemlock trees had been stripped of their tanbark and the timber harvested, many hillsides failed to support cultivation. But the area was relatively well suited to dairying. Cows flourished on the upland pastures, and farmers could send their products to New York City by rail.[5] Equally important, Broome County was home to the Endicott Johnson Shoe Company, which became the largest integrated boot and shoe manufacturer in the United States. E-J (as it was called locally) hired both rural migrants and recent immigrants to tend machines and perform other processes by hand for piece-rate wages.[6] Its vast factories spurred the rapid growth of Johnson City and Endicott, as well as Binghamton, the older commercial center at the confluence of the Susquehanna and Chenango rivers.

Broome County figured in national discussions of rural outmigration and social problems from the beginning because this region exemplified the rural depopulation and decline that worried conservative reformers. U.S. Secretary of Agriculture James Wilson remarked on the abandonment of farms during a tour of south-central New York in 1909, as did former president Theodore Roosevelt two years later.[7] The 1911 report of the Country Life Commission, which recommended remedial measures designed to make farming more profitable and rural communities more attractive,[8] ignored several factors whose importance was recognized by many country people and some agricultural experts at the time. The economic plight of rural residents did not arise solely from the limited incomes that even well-managed farms yielded. Lack of opportunity for young people to earn wages and accumulate savings that would allow them to marry and set up independent households, as well as dislike of the endless toil and scanty remuneration that farming entailed, propelled them to the cities. The exodus of youths, in turn, precipitated a crisis in the intergenerational transmission of farms. In the Nanticoke Valley, outmigration disrupted the customary process of farm succession. Although landowners could hire male and female help relatively cheaply until the Great War, they found that persuading their children to forgo a regular income and wait to inherit land in return for their labor was more difficult. In a 1919 article based on a study of Broome County farms, Edward G. Misner commented that "in the past ten years too many farm boys who would have made good farmers have gone into other higher paying industries, because they foresaw what wages they might expect in farming." Those who had gone to work in the city were much less likely than those working nearby to return to farm with their parents and eventually take over the enterprise, or to marry a son or daughter who would inherit the family farm. Older farmers said that the lack of successors was a major cause of their reluctance to expand their operations. Many elderly couples limped along as best they could but allowed the enterprise to contract as their strength declined. "The results are apparent," Misner said; "many farms are operated by men and women of advanced age; many others [are] for sale."[9]

The sale of farms might result in consolidation rather than the abandonment of productive land. The Youngs' story typifies the process; they bought the neighboring farm when they set up their partnership. From 1900 to 1940, the total number of farms in Broome County decreased steadily and the average size of farms rose slightly. In 1920, the regional supervisor of the agricultural census pointed out that the decline in the number of farms that had occurred since 1900 "was the result of permanent abandonment of farms in many instances," but it could be "partially explained by the combination of two or more farms into one."[10] With farm consolidation, agricultural production was maintained and productivity might increase.

Even the permanent abandonment of farmland in remote areas was not necessarily regrettable, agricultural economists contended. In 1928, Lawrence Vaughan found that "in south-central New York, the abandonment has been from 15 to 25%. In this region there are many scattered areas of abandoned farms...located at the higher elevations and...separated by the intervening valleys." The rate of decline in Broome County was typical of the adjacent counties. The poorest quality lands were abandoned first. In contrast to prewar analysts, who had found much of the land still being used as pasture by other farmers even though the houses were vacant, Vaughan found that about half the houses were occupied but the land was no longer used. He recommended that, if farm commodity prices ever revived, good lands should be improved rather than poor lands being reoccupied. A reforestation program would prevent these unproductive farms from "wasting the lives and money of new buyers."[11] After 1929, some landowners in Maine and Nanticoke reforested barren hillsides with stands of evergreens under a New York State program.

The use of land on Broome County farms shifted slowly but significantly from more intensive to more extensive uses and from the cultivation of crops to animal husbandry. This trend had emerged by the turn of the twentieth century and became quite marked by 1920. Although the average size of farms increased, the average amount of improved land on farms did not.[12] By 1930, half the farmland in the county was in pasture and only about a third was cultivated. In 1940, the statistically typical 113-acre farm would have had 40 acres in fields and meadows, growing feed and hay for cattle; 47 acres in pasture; and 26 acres of woodlot. About half the pastureland could be sown with improved grasses or converted back into meadow; the rest, which generally lay on hillsides, was permanent. The major trend from

1900 to 1940 was a steady expansion in the proportion of farmland devoted to pasture and an absolute decline in cropland. This shift reflected changes in the balance among different types of farms in the county and in the uses of land on individual farms.

"The Agricultural Progress of Fifty Years," a summary of changes in farming since 1850 included in the 1900 U.S. census report, explained the long-term trend toward the more extensive use of rural land in the North Atlantic states. "A portion of the decrease in improved farm land is due to the inclusion within city limits of former farm areas, but the greater proportion is due to the change in the character of agricultural operations, and the new methods adopted for securing the greatest income from farm land." Although the acreage in tillage decreased, "the growth of the city population in these states has stimulated certain special branches of agriculture, notably dairying and market gardening. These changes have led to a natural selection of land according to its adaptability to special uses." With the expansion of dairy farming,

> the most fertile and easily tilled lands have been retained under cultivation or have been converted into permanent meadows and made increasingly productive, while less fertile lands that are plowed with difficulty, and meadow land which can not be mowed by machines, have been, in many cases, converted into permanent pastures. The resulting increased average fertility of plow and meadow lands enables the farmers to raise on a small area the winter feed for the animals that can be kept on the enlarged area of partially exhausted pasture land during the summer. The increasing cultivation of forage crops, the use of the silo, and the larger acreage of corn grown and fed on the farm, are all factors contributing to the same end—a decrease in the total area required to produce the winter feed for the farm animals. No such improvements have been made in the pasture lands; hence, there is a readjustment of the total farm area, involving a reduction of meadow and plow land and an increase in that used for pasture. The tendency toward this change—arising from the increased average productiveness of the soil still under plow or mower—is enhanced by the custom, growing among Eastern farmers, of purchasing feed produced in the West. This practice lessens the demand for meadow and plow land, and results in an increase in the area used for pasture, so that a greater proportion of farm land in each year being [sic] considered as unimproved.[13]

Officials were anxious to reassure the public that the decline in improved acreage was not the result of the wholesale abandonment of land and did not involve any real decline in farm productivity.

In agricultural economists' eyes, the shift from tillage to meadow and from meadow to pasture, rather than being a response to soil depletion, was a successful adaptation to changing economic conditions. Specialized dairying, while requiring less tillage and more pasturage than the mixed farming that had preceded it, maximized the productivity of land and involved an increase in the scale of farm operations. Older attitudes favoring arable over pastoral farming lingered on, expressed in the enthusiasm that the report showed for market gardening in the vicinity of cities, but a "scientific"—or, rather, capitalist—attitude that regarded productivity and profit as the crucial criteria for evaluating farming systems was clearly in control. Within the evolutionary social-scientific mode of thought that framed the report, the expansion of specialized dairying was progressive, even though it was accompanied by a decline in the intensity of land use.

This formula for financial success lay beyond the reach of most Broome County farmers. The conversion of old fields to meadows and of hilly hayfields to permanent pastures was not accompanied by any increase in productivity. Few farmers, including the Youngs, had enough fertile well-drained land to grow sufficient corn to feed their dairy herds through the winter even though they adopted ensilage. Most farmers were too poor to afford these investments, and their land was ill suited for the system of capital-intensive farming that the U.S. Department of Agriculture (USDA) and agricultural experts recommended. Changing patterns of land use did not yield the prosperity that might be enjoyed by farmers with better land and greater operating capital.

Evaluating Alternative Approaches to Agriculture

What, then, would improve the productivity and profitability of farms that were not on submarginal land that should simply be reforested? A series of studies by experts at the New York State Agricultural Experiment Station in Geneva and the New York State College of Agriculture in Ithaca, just northwest of the Nanticoke Valley in Tompkins County, was designed to ascertain which types of farming were most successful in this region and used Broome County to illustrate the difficulties that upland farmers faced. Begun in 1908 under the supervision of George F. Warren,[14] these studies asked typically capitalist questions about investment and incomes and applied regular business accounting procedures to farms, calculating their expenses, net incomes, labor incomes, and returns on investment from questionnaires completed by farmers.[15] But the statistical results for upland farms did not demonstrate the benefits of a capitalistic management strategy. Contradicting experts' assumptions, they demonstrated the rationality of approaches that limited expenses for feed and relied on family rather than hired labor.

Study participants were solicited from the membership of farm organizations, which meant that more successful farms were overrepresented.[16] Strikingly, dairy farms were not more profitable than diversified farms; their higher expenses for feed and labor offset their higher returns. Although diversified farms yielded more profits as investment increased, on dairy farms the key to success was limiting costs. Male farmers reported that unpaid family members—chiefly their wives and children—performed 21 percent of the value of all farm labor. On larger-scale, specialized dairy farms, women did 18 percent of the total value of labor, and many families had a hired hand year-round. On diversified farms, women performed 26 percent of the total value of labor, and expenditures for hired help were lower.[17]

Designed "to ascertain…why some farms pay better than others," this study ended up probing the reasons why so many farms in this upland region did not pay at all. Even valley farms were only marginally profitable, and labor incomes from the hill farms were just half of those from valley farms. Only one-third of the male farmers who participated in the survey earned labor incomes that were higher than the wages of a hired hand, which was less than a dollar a day. Some reaped only the "living" their land and labor yielded: the dairy and poultry products, beef and pork, vegetables, and fruit that they consumed, as well as the wood that heated their houses. Farmers' fundamental difficulties, as the economists saw the situation, were lack of capital and inefficient use of labor. Agricultural economists criticized farmers' management methods. Fully 45 percent kept no systematic accounts. Those who recorded expenditures and receipts from sales did not keep track of their own and others' unpaid work, and taking labor time for granted made it impossible to evaluate the efficiency of their operations. "Much of the labor is not paid, and the inference has been drawn that time is of no value."[18]

Strikingly, agricultural economists exhibited a similar attitude when it came to the value of women's labor. They acknowledged that "the greater part of the work on farms is done by the farmer and his family"; unpaid family labor amounted to over 70 percent of all labor on these farms. Yet they pronounced employing a hired man and enlarging the operation to utilize his labor efficiently the key to profitable farming. The way they estimated labor costs deserves scrutiny. "Farmers were asked to estimate what the work they did would cost" if they had to hire a man to replace themselves and other family members. The basis on which men estimated the value of their wives' and children's labor is never explained, however. Curiously, the economists said little about their data showing that, as the proportion of unpaid family labor increased, the farmer's labor income rose substantially.[19] Instead, the authors opined that "very few women…help with the milking." Yet their specific comments contradict this assertion. They acknowledged that "in some parts of the State where nothing but milk is sold, the women commonly help milk. This is a recognition…that dairying provides only a partial day's work. A woman will milk as many cows as a man and do the housework besides. In this case a woman saves the wages and board of a hired man….The objection to hiring a man is that the system of farming does not provide a full day's work for him. When a man and his wife both milk, it is possible to make a living and gradually pay for a farm."[20]

Figure 5. "Stock parade, Bath Fair," New York, September 1945. Mother and child show cow and calf. *(Photograph by Charlotte Brooks, Standard Oil (New Jersey) collection, University of Louisville Photographic Archives. [SONJ 30501])*

Warren's summary of these surveys observes that larger-scale farms might profitably employ an adult son and that their more substantial returns might induce a son to remain rather than move to the city.[21] Although this interpretation identifies a higher labor income as a cause rather than an effect of an intergenerational partnership, it is more likely that the relationship was reciprocal and the unpaid labor of an adult son or two might well subsidize farm expansion, as happened in the Young family.

A 1915 report on milk production in Delaware County, located just to the east of Broome County, analyzed dairies milking at least a dozen cows. These farms led the state in fluid milk production, in part because soils in this region had ample lime and good drainage. Yet the total costs of milk production exceeded the receipts by 32 cents per hundredweight. How, the report asked, was it possible for farmers to survive under these conditions? Most accepted a very low rate of return on the labor they performed. Computing labor costs was difficult, the author acknowledged. "Man labor" was estimated at 15 cents per hour, but "there were no records available to show how much women and child labor costs. In this investigation, women and child labor was valued at 10 cents per hour," while "horse labor" was valued at 12 cents! Women and children's work made up 19 percent of all the labor in milk production. Detailed data on "farms using women or child labor to care for cows" showed that women cared for cows on 60 percent of farms, children did so on 6 percent, and both women and children on another 19 percent, so 86 percent of all these farms utilized the labor of women and children in the barn. "When two or more family members work, it is possible to live on a farm and receive lower wages than" the experts' estimates. "The average farmer, by working hard himself and getting help from his wife and children, is able to live on the land." The labor that women and children performed was crucial to farm survival.[22]

E. G. Misner's "An Economic Study of Dairying on 149 Farms in Broome County, New York" faced these uncomfortable facts directly. He conducted this survey in 1915 with the cooperation of the Broome County Farm Improvement Association, to which George Young belonged, and with assistance from the county agent, E. R. Minns. Data were collected on milk sold to processing companies by farms with six or more cows, so larger-scale operations were overrepresented. Misner recognized the difficult environmental conditions in this region: "Although Broome County may be considered a leading dairy county of the State, many farmers, especially those on the uplands, follow an extensive system. An abundance of pasture, a short growing season, and a soil that is not naturally fertile, encourage the summer system."[23]

In evaluating labor costs, Misner adopted the rates used in the Delaware County study. Adult male labor was charged at the average wage earned by a hired man, 15 cents per hour. Misner declared that because "the time of women and children usually is not so valuable as the time of men...it was charged at 10 cents an hour." A detailed table of the labor used in specific tasks shows that women milked on one-quarter of these farms and were even more often involved in caring for the cows, making dairy products, and cleaning dairy utensils, including mechanical separators. Women performed 19 percent of the total labor in the dairy.[24]

The analysis of production costs and receipts from sales showed that most Broome County farmers were losing money on their dairy operations. Only 61 of the 149 farms showed a profit. The Youngs were not among this fortunate 41 percent; their dairy was not profitable until Ralph and Stacey bought more land and the two generations went into partnership. The question then arises: "How do such producers remain in business? The answer is that they...accept lower wages than the rate at which their time is charged"; the average rate of return for human labor on these farms was just 11 cents per hour. Misner was predisposed to regard larger herds as more profitable than smaller ones. His detailed figures, however, showed that farmers lost more money per cow when they milked 11–17 cows than they did when they milked 6–10. Similarly, summer dairying was not as unprofitable as winter dairying. Farms "back in the hills" had lower costs, although their cows were just as productive. "Practically all of the costs are higher when a larger proportion of the milk is produced in winter." Although milk prices were also higher, the costs for feed and labor more than made up the difference.[25]

Significantly, Misner found that "a larger proportion of the labor is done by women and children on summer than winter dairies." "Since this woman and child labor was charged at a lower rate than man labor, it tends to reduce the cost of labor for the summer herds. By thus working the whole family, it is often possible to save on the expense of a hired man."[26] Like Warren, Misner recognized that reliance on unpaid family labor was crucial to farm survival. Even when women's labor was devalued, as it was in all these studies, the figures showed that women made a considerable contribution to the labor that dairying required. Saving the expense of a hired man, who might well earn as much as the owner-operator, represented an even more substantial contribution to the family's income. The "two-man dairy" that agricultural economists regarded as optimal in this region might well be a husband-wife operation.

Between 1919 and 1933. Misner conducted a series of economic studies of dairying on farms in central New York, analyzing the changes that accompanied the shift from making butter on the farm to selling cream and then fluid milk. Making an important modification in his methodology, he asked farmers to estimate the value of family members' labor rather than assigning it an arbitrary—and, for women, a blatantly low—value. "The operators were asked to state a wage per hour that would represent what it would cost to hire a person equally efficient to do the work they were doing, and also to give rates for other classes of help." Strikingly, adult male farmers attributed significantly greater value to the work of their wives and children than the agricultural economists had done. On average, farm operators valued their own work at 41 cents per hour and their wife's work at 32 cents per hour, representing a ratio of 4:3 rather than the 3:2 ratio that economists had used. Equally important, most men valued their wife's work more highly than that of a hired man. Indeed, one-third of farmers valued their wife's labor as highly as they did their own, worth between 35 and 50 cents per hour.[27]

In a report based on statistics for 1919, Misner documented women's substantial contribution to the dairy operation. Family members did 80 percent of the total labor; male farm operators reported that they did 52 percent and their wives 12 percent. Detailed analysis of the participation of farmers' wives in milking and the care of cows, dairy products, and utensils shows that wives performed a substantial part of the milking, putting in 27 percent of the total hours during the summer and 21 percent during the winter. The hours women spent on washing up milk utensils were relatively less important than they had been in 1915, but the hours they spent on milking had markedly increased. The wartime shortage of labor made women's participation even more essential.[28] After the war, Misner compared different ways of marketing milk and organizing labor. Strikingly, more than two-thirds of the male farmers he surveyed reported that their wives did a considerable share of the work and evaluated their labor as at least as valuable as that of their sons and hired men. In the more intensive operations, however, just one-quarter of wives were reported as working on the farm, although those who did performed labor worth just as much as that done by sons and hired hands.

P. H. Stephens analyzed the keys to successful dairy farm management during the years 1921–1925 for all the types of farms covered in these studies. Although he, too, was predisposed to favor intensive winter dairying, he found that under most conditions farmers did better by avoiding high feed costs and relying mainly on summer pasture. They also gained by utilizing family rather than hired labor. Stephens pointed out that economic conditions required farms to increase in scale merely to survive:

> There is an increasingly greater gap between the efficient and inefficient producer. Whereas under a system of farming...approaching the self-sufficing type, inefficiency resulted merely in a lower standard of living, under the present day commercial type of agriculture, the inefficient are being deprived of the very means of production. With the increase in the use and application of scientific knowledge, the invention and adoption of modern machinery, the increased capitalization of our agricultural resources, and the growing complexity of our marketing problems...this situation is both the hope and the despair of the agriculturist of the future. For the competent man who has the means of acquiring a foothold, farming will offer greater financial inducements in the future than it has in the past. For the man not so equipped, farming may become less attractive as the full effects of the present evolution in agriculture become more evident.

Despite this talk of efficiency, the capital resources that farmers had at their command were more crucial to the profitability of their enterprises than their skill in managing land, livestock, and labor. Stephens predicted that farmers with little capital would go out of business or be unable to get started, while successful farmers would have to act like other businessmen.[29]

That fate was delayed, although not entirely averted, by the economic depression that began in rural America after World War I and engulfed the rest of the nation after the 1929 financial crash. Instead, Broome County agriculture exhibited an increasing divergence between large-scale, specialized farms and small-scale, more diversified ones.

Changing Patterns of Farming

Dairying remained ubiquitous on Broome County farms and expanded through continuous growth on a broad base. The average number of milk cows on farms that had cows, which is the best indicator of scale, rose steadily from 7.1 in 1900 to 11.5 in 1940. Just over one-sixth of these farms milked 20 or more, but more than half still milked fewer than 10. Nanticoke Valley farmers agreed that a dozen cows was about the most that a family with two experienced adults could manage without the twice-daily assistance of additional family members or a year-round hired hand. Hiring outside help for the dairy was regarded as a dubious proposition because only family members could be trusted to work faithfully and had enough of a stake in the welfare of the herd to do the job well. Most farms milking 20 or more cows were, like the Youngs' operation, conducted as partnerships between parents and adult children; a few older farmers, like the Briggses, hired another family, who lived in a small house on the property. By 1940, a significant fraction of Broome County farms had crossed that size threshold, but the majority continued to operate on a scale that a husband and wife or parents and children could handle.

Farms milking fewer than five cows generally made butter or sold cream to condensaries rather than selling fluid milk. The connection between the scale of dairy operations and the form in which dairy products were marketed solidified as processing and marketing were consolidated. In the stories that Nanticoke Valley farmers told, enlarging their dairy operation enough to ship fluid milk was an achievement that took years or even decades of effort, expanding and improving the herd by breeding or buying better heifers and reinvesting the butter proceeds or creamery checks in enlarging the barn and purchasing machinery. Many who tried to make this transition were unable to do so.

Poultry production was even more widespread than dairying and expanded almost as rapidly. The report on the 1900 agricultural census remarked, "It is only within comparatively recent years that the production of poultry and poultry products has assumed the proportions of a distinct industry. It was, and to a decreased extent is yet, a sort of collateral undertaking, or mere incident in general farming, conducted by the farmer's wife. With but little attention given to the welfare of fowls, the results are often meagre and unsatisfactory, but when intelligently conducted there is probably no branch of animal industry from which are secured such quick returns of money invested."[30] The gendered assumptions underlying this statement, especially that a woman who kept chickens was not a farmer but merely "the farmer's wife," were not shared by the rural residents of Broome County. Outside experts recommended that men take over the poultry operation, increase its scale dramatically, apply scientific knowledge and industrial techniques, and turn it into an efficient business enterprise. Across Broome County, however, most of the farm families who enlarged their flocks and earned more money from the sale of eggs and chickens did not adopt the capital-intensive, male-dominated model that experts advocated. They built better chicken coops and put nests in hen houses, administered remedies for mites and other common diseases, and purchased feed to supplement the grains they grew. But in most families, including the Youngs, women continued to take primary responsibility for farm flocks as they expanded.

The compatibility of poultry raising with commercial dairying as long as it remained primarily in women's hands was clear to some agricultural economists. In his analysis of farms selling Grade B milk in the early 1920s, Johann C. Neethling emphasized that "a farm flock always fits into the farm business" and "during the agricultural depression poultry was relatively more profitable than other enterprises." His conclusion vindicated the practices followed by Broome County farmers: "The poultry flock can be cared for by the women and children and need not interfere with other enterprises. During the spring and summer months when crops require most attention, the cows can be turned out to pasture and the work of caring for them is reduced."[31] The combination of summer dairying and poultry raising was less risky and potentially more profitable than specialization in either one.

Most farms kept poultry flocks and sold eggs and chickens. A few specialized poultry farms, like that operated by Venley McGregor and his family in Maine, used modern equipment and scientifically tested techniques to hatch large numbers of chicks that they sold to rural women with small flocks. On their own modest acreage, the McGregors kept hundreds upon hundreds of pullets in heated, lighted shelters year-round; fed them well to maximize egg production; and hired local women to work part-time grading and packing eggs for sale. When their own production fell short of what their contracts required them to deliver to city stores, the McGregors bought eggs from the farm women they supplied with chicks. Large-scale poultry producers did not compete with those who kept farm flocks; the two had a symbiotic relationship.

Farmers shifted the mix of crops they cultivated as they concentrated on growing more feed for their dairy cattle. Oats were by far the most important grain. Farmers who, like the Youngs, limed their fields harvested considerable quantities and had their oats ground together with purchased supplements. Corn was also regarded as a valuable part of dairy cows' diet, particularly when the whole plant, ears, stalks, and all, was chopped in the field and fermented in a silo. Under optimal conditions its yield per acre and nutritional value per bushel were comparatively high. But conditions in Broome County were far from optimal. Early frosts often prevented the ears from ripening. The heavy clay and silt soils of the upland areas were too wet and much too acidic for corn to thrive. Although the richer soils in the valleys were more hospitable, in some years the cornfields there were flooded, which was fatal to the crop. Farmers in the Nanticoke Valley said that when the corn was puny because the spring was too cold and wet and the summer too hot and dry, they could expect a bumper crop of hay. Corn cultivation remained patchy and small scale; the average amount of land planted in corn in Broome Country declined steadily from 1900 to 1930, and by 1940 just half the farmers grew any corn at all.

Indeed, as late as 1940 more land was planted in buckwheat than in corn. This small grain was much less nutritious and yielded only a few bushels per acre, but it would grow almost anywhere and could even be planted after other crops had died. Soil scientists advised farmers to sow marginal land in buckwheat, plow the grain under when it was green, and convert the old field to pasture. Yet buckwheat held its place in many Nanticoke Valley farmers' survival strategies. Although some farmers on the valley floor scorned those on the hills whose buckwheat was "no higher than the bees' knees" in mid-July, most understood that otherwise those fields would yield nothing but burdock and thistles. Buckwheat was decent feed for cows that were dried out over the winter and the key ingredient of locals' favorite pancakes. Pitcher's Mill near Maine village not only ground it on a custom basis but also purchased some for its popular Sunrise Pancake Flour.

Farmers were in a much better position to cultivate hay and forage crops. In 1900, Broome County farms had, on average, 24 acres in hay, most of it timothy and clover. Over time the amount of land devoted to hay increased, and by 1940 farms produced on average more than 40 tons of hay and forage annually. A few farmers, like the Youngs, experimented with alfalfa and field peas, legumes that were recommended for their high yield and nutritional value as well as for enhancing soil fertility, but neither could be grown by itself on local soils without the liberal and expensive application of lime and chemical fertilizers, which defeated the point of planting them. Dairy farmers had to purchase considerable amounts of feed for their stock. As early as 1910, over 80 percent of farmers reported spending money on feed, which constituted their largest single operating expense. Over the next thirty years, the amount that farmers spent on feed tripled, far exceeding the cost of labor hired for haying and harvest.

Most rural residents, including those who did not farm, raised vegetables and potatoes for their own use, and those who devoted their energies to market gardening might earn a considerable sum each season. The proximity of the public markets in Endicott, Johnson City, and Binghamton enabled people in the Nanticoke Valley to profit from such small-scale operations. Large-scale commercial operations were rare, however; greenhouses were expensive to construct, and it was hard to find stoop labor when better-paying factory jobs were readily available. Although most farmers had orchards and some sold small quantities of fruit and apple cider, the uplands in Broome County were less well suited to commercial orchards than were other parts of south-central New York. Only a few farmers, such as the Green family in East Maine, specialized in fruit-raising and conducted large-scale enterprises.

Many Nanticoke Valley farmers were in precarious circumstances through the 1920s, and economic conditions deteriorated further during the 1930s. What is most striking about the detailed information compiled by the agricultural censuses on the scale and specialization of farms during these difficult decades is the persistence of medium-size, diversified farms and the proliferation of very small ones. If one adult in a family devoted his or her efforts to subsistence and small-scale, market-oriented operations, the value of his her products was about equivalent to what a semi-skilled and experienced adult male could earn by working full-time in the boot and shoe factory. The Wrights, whose farm was more typical than the Youngs', followed a different strategy to maintain their income. Jerry Wright had inherited land, but the family had less available labor than the Youngs did. Their dairy remained somewhat smaller, with about a dozen cows that Jerry and Ernest, his adult son, handled by themselves; Jerry peddled butter, buttermilk, and cheese in nearby Johnson City. Their enterprise also remained more diversified; they sold potatoes and cabbage, apples and cider, and even maple sugar in the city as well.

The 1930 and 1940 federal censuses of agriculture shed light on the relationship between the degree of specialization and the scale of farm operations. In 1930, "general," "self-sufficing," and "part-time" farms made up 35 percent of all farms in Broome County. They averaged 66 acres, so their small scale was less a matter of available land than of working capital. Almost all sold dairy and poultry products, and most provided an adequate subsistence. The most unusual and intriguing are the "part-time" farms "where the [male] owner spent 150 days or more at work for pay at jobs not connected with his farm, or reported an occupation other than farmer."[32] These farms, which were operated by families in which the adult man held an off-farm job, made up at least 15 percent of all farms the county in 1930. In the Nanticoke Valley, these farms were operated both by long-time residents who lacked the resources required to make a living on the land but were unwilling to give up and move to town and by newcomers, many from immigrant backgrounds, who bought farms from native-born families who had decided to depart. The combination of farming with wage-earning was more pervasive than the census definition indicates because in as many families the husband worked on the farm while the wife and older children worked in the factory. Subsistence production was vital during the Depression; in 1940, the value of the products devoted to household use was greater than the value of any group of products sold on 37 percent of all farms. Trends toward increases in scale and specialization were halted and, perhaps, even reversed, indicating that many families lacked the land, labor, and operating capital required to expand their dairy operation. Indeed, the proportion of farms that specialized in dairying actually declined from 52 to 45 percent.[33] Many diversified farms had small dairies, but most had been unable to make the transition to large-scale operations. Lack of capital, including land that was adequate to grow feed, coupled with the decline in commodity prices that limited the income that farmers received, prevented them from expanding, much less embarking on more capital-intensive forms of production. The decline in off-farm employment made families' immediate subsistence needs even more compelling. Farmers focused their efforts on providing milk, meat, potatoes, vegetables, fruits, and fuel for themselves, hay and fodder for their cows, and grain for their hens. Small-scale, labor-intensive, diversified production was the only strategy readily available to them if they wanted to hold onto their land. But that yielded only a modest and seasonal income. Although it might go a long way toward paying the grocery bill or the taxes, it came nowhere near what was required for expansion.

At the same time, the scale of specialized dairy operations continued to increase. This trend was the paradoxical result of hard times. As the smaller and less profitable dairy farms found it harder to stay in business, those that survived were larger in scale and more specialized than most dairy farms had been in 1930. They were also more distinct from the other farms in the county. In 1940, the average total value of their products was one and two-thirds that of all farms. Indeed, the scale of specialized dairy farms was so large that, although they made up only 56 percent of all farms with dairies, they produced fully 96 percent of the total value of dairy products sold in the county.

Hiring Labor or Taking Off-Farm Employment

By 1940, Broome County farms were clearly divided into two classes by their labor systems: large-scale, specialized farms that hired some labor, and small-scale, diversified farms sustained by families who often combined wage-earning with farm work. This class polarization differed profoundly from the situation in 1900, when farms had been distributed along a continuum. To be sure, there were wide disparities in the amount of productive property farmers owned and in their standard of living. Families with hilly land, relatively little capital, and low cash incomes were in a very different position from those with ample land near the creek and incomes that allowed some investment in improvements. But both better-endowed and disadvantaged farmers faced essentially the same conditions and occupied structurally similar positions in relation to merchants and shippers, especially milk processors and dealers. Comfortable and impoverished families also had common goals: to provide for their own subsistence; secure their ownership of the land; and, if possible, increase the value of the farm. They were interested in expanding the scale of their enterprise mostly to help their children get started and provide for farm succession. In this rural society, the sharpest class division was between families who owned productive property and those who had no land, not between better-off and poorer farmers.

The expansion of capitalist relations that occurred in the agricultural economy during the early twentieth century fundamentally transformed the structure of rural society. Both large-scale, specialized farms and small-scale, diversified ones sold products in urban markets, but they put together a livelihood in strikingly different ways. Large-scale enterprises were operated by full-time farmers; although they relied on family labor, they also employed hired hands. Small-scale farms were generally operated by farmers who worked for wages themselves, occasionally on other farms but more commonly in factories in nearby cities. The emergence of this class division is revealed by the statistics for farm and off-farm labor. Over time, the proportion of farms that hired labor declined, but the average amount of labor they hired increased. Simultaneously, the proportion of farmers who worked for wages rose to a substantial level. Contrary to the popular image of farmers as independent producers, just one-third of Broome County farmers were neither employers of labor nor wage laborers by 1940.

In 1910, almost two-thirds of all farms reported paying wages to farm labor, but most hired hands worked during planting, haying, and harvest. By 1940, only 34 percent of all farms reported paying any money for labor, but their average expenditure was more than twice the average value of the products consumed by the household on farms in the county. In 1940, 83 percent of all farms reported that an average of 1.6 family members were working on the farm. Three-fifths of all farms used only family labor. But the larger-scale, more specialized farms that had hired labor had an average of 1.6 nonfamily workers each, which represented a doubling of the labor force. One-sixth had full-time workers throughout the agricultural year and additional hands to help with seasonal field work.

Over time, farmers became more likely to work for wages themselves. In 1940, nearly two-fifths of all farmers had worked off their own farm during the previous year. Strikingly, they averaged 187 days of paid employment, just over half of a farmer's year and over two-thirds of the days worked by full-time employees in nonfarm occupations. Only 5 percent of farm owner-operators had worked for other farmers, most often in haying and harvest. Those who had done nonfarm work, by contrast, averaged 198 days of paid labor the previous year. The agricultural economy and rural society were closely enough integrated into the capitalist system that large-scale, specialized farms had assumed the character of business enterprises and small farms depended on waged employment as well as on diversified production.

The specialized dairy farms that dominated Broome County in 1940 exemplify the consequences of agribusiness. Changes in the processing and marketing of dairy products locked farmers into the characteristic cycle of capitalist production; they had to continually expand their operations simply to stay in business. They adopted innovations in feeding and milking their cattle that were supposed to reduce their unit costs and their labor, but the marginal and decreasing amounts of money and time they saved were devoted to increasing production. Milking more than a dozen cows year-round required farmers to buy feed, raise and store corn silage, improve their barns, and purchase milking machines, as well as hire labor. Expansion, rather than being a way to raise incomes, was a precondition of viability. Eventually its capital costs became too burdensome for most farmers. Although the dénouement did not come until after World War II, when dairy farmers had to maintain much larger herds and buy bulk tanks in order to sell milk at all, the causes of the eventual contraction of the industry are visible in the interwar period.

By 1940, these farms were highly specialized, capital-intensive, and market-driven enterprises. Under capitalist conditions, increases in scale transformed these farms qualitatively as well as quantitatively. Most were still family-owned and -operated, but their viability depended almost entirely on market forces. To provide for farm succession, farmers had to increase their profits enough to sustain two generations. These family farmers were no longer simple commodity producers that reproduced themselves outside of capitalist economic relations; to perpetuate the farm, owners had to set up legal contracts ensuring adequate provision for the older couple, fairness in the division of the capital among all the heirs, and compensation for the years of unpaid labor that the successor had invested in the operation. Securing financing almost always required a mortgage on the land, which in turn mandated increases in scale and additional investments in livestock, buildings, and machinery. All that kept capitalist relations from penetrating agriculture's very core were farmers' family commitments, their immersion in rural culture, and their opposition to the dominant political economy.

This analysis emphasizes the loss of control over farm decision making that families experienced as they became more involved in capitalist production. There were two primary contradictions in their situation. First, in contrast to most business firms, they did not control the processing and marketing of their products. Rather, they operated within a commodity market that was to a large extent controlled by corporate interests that extracted their profits from primary producers. This contradiction led large-scale farmers to organize cooperatives. Second, no matter how involved they were in capitalist markets, family farms were not capitalist firms. They were often forced to behave as if they were, but to most farm families that felt like a strain, or even a violation of their fundamental values. Relationships between parents and children and between husbands and wives were subjected to structural tensions when labor arrangements and family decisions were controlled by financial considerations. Women and children occupied different positions relative to productive property than men did, and when capitalist economic relations penetrated family farms the contradictions inherent in both gender and intergenerational relations were exacerbated. At times, it seemed that the interests and demands of the farm were in conflict with those of the family. Most farm people, women as well as men, responded by asserting the indissoluble unity of family and farm, but that did not solve the problem; rather, it often concealed a struggle over the definition of the interests of the farm and who had the power to speak for the family.[34]

These relatively successful farms occupied a structurally different position in the rural economy than small-scale, diversified farms did. In the 1930s, although a few families still struggled to increase the scale of their operations, that was no longer really possible. At best, they maintained themselves and their farms by producing eggs, chickens, vegetables, fruits, and berries for sale in nearby urban areas. In Broome County, unlike most rural areas across the country, many families with marginal enterprises were able to remain on the land by engaging in wage labor. Their earnings took the place of the cash income that other farmers obtained from their commercial operations, supporting the family and sometimes subsidizing the farm itself.

When the small-scale, diversified farms that remained common in Broome County were analyzed within the cost-profit framework applied to capitalist enterprises, they appeared to be losing propositions. Most farmers did not earn any labor income; within a capitalist calculus, the time they spent doing farm work did not pay. What they sought, instead, was the security of knowing that they would not go hungry even in the midst of the Great Depression and the independence of being able to produce something for themselves rather than relying entirely on an employer.[35] In their own terms, they were working people who sought a degree of independence from capitalist-controlled wage labor and the market for consumer goods. Although their farm operations did not allow for expansion, they enabled these families to sustain themselves.

Class, Labor, and Gender

The social consequences of the stratification of rural landowners into two distinct classes have often been overlooked, largely because historians and social critics have been more concerned about landlessness and tenancy. In part, the division was reinforced by geography. In this hill-and-valley area, farms were vertically stratified. Most of the capital-intensive operations were located on the valley floor, while those on the hills were more land-extensive. But landowning families who combined farming with wage-earning lived in the valleys as well as back in the hills. The fact that many working-class landowning families were foreign-born mattered to some degree, but many longtime local residents resorted to the same combination of farming and wage-earning to hold onto their land.

The fundamental division between these two groups of farmers sharpened throughout the period before World War II and had significant effects on rural institutions. While farmers with larger-scale, specialized operations organized purchasing and marketing cooperatives to gain leverage in the market, farmers with smaller-scale, diversified operations who were also wage-earners more often processed their products themselves and sold directly to urban consumers, sometimes at their workplaces or in their former neighborhoods. The Dairymen's League, Grange League Federation, and other cooperatives no longer included the majority of farmers and came to resemble agribusiness corporations.[36] Both classes conducted their lives on a regional rather than local scale; specialized farmers allied with their counterparts across the county and state, and working-class families' economic and kinship ties bridged the urban-rural divide. The fundamental difference in their approaches to labor meant that gender relations were also inextricably connected with class position. In poorer families, women's labor on and off the farm was recognized as essential to survival. On larger-scale, specialized farms that hired labor, women's work was more readily marginalized.

PART III

THE DIVISION OF LABOR
AND RELATIONS OF POWER

5

Sharing and Dividing Farm Work

In rural societies, social relations are constituted around work. Relationships among women and men in farm families are shaped by the ways in which daily and seasonal work is organized. When husband and wife work side by side in the main income-producing operation, such as dairying or poultry raising, they interact frequently on a common terrain. Each spouse is well aware of the other's contribution to the family enterprise and is well informed about whatever problems arise: when the corn is stunted by too much rain, when the supply of feed might not last the winter, or when the milk check is too low to cover the cost of keeping the cows. They act as partners in a small business, as well as parents in the family. Sharing labor makes their relationship very different from that of couples who divide tasks sharply by sex, especially if the man is solely responsible for work in the fields and barn and the woman is confined to the house and garden. In *Bonds of Community*, I showed that farm families in the Nanticoke Valley during the late nineteenth century exhibited great flexibility in the gender division of labor.[1] Women and men generally worked together in the barn and hayfields and organized the rest of their labor in an amazing variety of ways, depending on their personal preferences and the composition of their households. Although women ventured outdoors to help their husbands and fathers more often than men assisted their wives and mothers with indoor tasks, all placed the highest priority on dairying, the main income-producing operation on most farms.[2] Women themselves regarded housework as less important and, often, less enjoyable than productive labor, such as vegetable gardening, poultry raising, and maple sugaring.

Labor also links families, which are organized around gender and generational relationships, within networks of sharing and exchange based on kinship and neighborhood. People interacted as members of families and kin groups, pooling labor and changing work to complete such seasonal tasks as haying and the harvest on adjacent farms. Even in such routine economic acts as hiring extra hands and marketing produce, people engaged with others through their familial and neighborly connections. Any person who conducted his or her business as a solitary individual would be regarded as an object of pity rather than as a powerful social actor; isolation from family deprived people of access to essential resources and support networks. Families' positions within rural society were dictated by the productive property they owned, the labor they could mobilize, and the strength of their social connections, not just the financial resources they controlled. The centrality of gender and generational relations to this nexus has placed farm families at the center of analyses of the capitalist transition, just as historically it positioned them—along with urban working-class families—on the front lines of economic change.[3]

Figuring out how gender figured in this great transformation has long puzzled students of history and political economy. Over time, the ways this question has been formulated have shifted from asking how capitalism affected women's position in the household and the labor market to how women of different classes and racial-ethnic groups participated in or resisted the social changes wrought by the extension of capitalist relations into more and more aspects of life. Feminist historians soon realized that the household and the labor market are not separate domains, but are both shaped by patterned, asymmetrical, and often hierarchical relationships among women and men. The authority that a man exercised within his family, for example, depended on the advantages he enjoyed in the workplace and the market, his recognition as the head of his household in civil society, and his rights as an active citizen. Women's subordination was manifest in sex-segregated jobs and sex-linked wage differentials; in the countryside as well as the city, women were paid no more than half of what men earned and toiled in the least secure, most laborious jobs. In the legal system, women's identities were subsumed by their fathers, brothers, husbands, and sons. Many married women's names were not on the deeds to the land. Even when they inherited property from their parents or late husbands, they seldom controlled it independently. Until 1920, women were excluded from active citizenship, unable to represent themselves in electoral politics except in school district affairs.

Feminist scholars in the materialist tradition sought the key to women's position in gendered relationships to the means of production, but recognized that women's work in reproduction—both intergenerational reproduction through childbearing and childrearing, and the daily reproduction of labor power accomplished through housework—shaped their lives as

powerfully as their paid employment and lack of property did. In historical perspective, it was not at all clear why, when capitalist industrialization
shifted income-producing work from the household to the factory, household labor itself became devalued, even though it remained essential. Was
any set of tasks that women performed automatically deemed less valuable
than tasks customarily performed by men? What connections were there
between women's primary responsibility for childcare and their relegation
to a secondary, often casual labor market?

Models based on the experiences of the urban middle and working
classes during the process of industrialization were seldom applicable to
rural women. The first generation of women's historians discovered that
farm women's lives had to be examined through a different lens because production and reproduction, kinship and property remained interconnected
rather than being sundered in the capitalist transition. Farm women continued to make essential contributions to their families' livelihoods through the
mid-twentieth century, producing a subsistence by gardening and canning;
earning income through "sidelines" such as poultry flocks; and participating
actively in the main income-producing operations, whether milking cows
and raking hay or feeding hogs and driving tractor in the cornfields. But the
value of women's labor did not necessarily translate into authority in farm
family decision making. Even though men recognized that women's work
was essential for farm survival, many refused to treat their wives, sisters,
and daughters as full partners. Farm families, like urban families whose
members held disparate positions in the labor market and civil society, were
sometimes riven by overt or covert power struggles.[4]

In her path-breaking book, *Preserving the Family Farm: Women, Community, and the Foundations of Agribusiness in the Midwest, 1900–1940*,
Mary Neth sidestepped false dichotomies between capitalism and patri-
archy, public and private, and property and labor, demonstrating that the
patriarchal and privatizing forces that accompanied capitalist agriculture
were countered by reciprocal and communal ties maintained by families and
neighborhoods, often in self-conscious, collective opposition to market pressures. In *Bonds of Community*, I argued that women played a central role in
the construction and continuation of mutuality, a culture of reciprocity that
shaped interactions both within and among farm families. Neth adopted
this idea and drew out its political-economic implications. In her analysis
of the rise of agribusiness, Neth paid careful attention to the thoughts and
actions of farm people, not just to the statements of policymakers and the
designs of agricultural institutions. By "placing gender, family, and community at the center of the story and using them as a lens to reexamine agricultural history," Neth was able to rescue the forgotten alternative to capitalist
farming and its hierarchical structures of class and gender.[5]

As Neth traced its emergence in the early-twentieth-century Midwest,
modern agriculture, which she called agribusiness, was capital intensive,

maximizing production through the purchase of inputs, chiefly fertilizer and feed, and the mechanization of labor. Specialized farms shipped commodities to processors that sold them on national and international markets. Agribusiness undermined the relations of production that were customary in rural communities, especially the sharing of labor and conservation of cash. Many farm families lacked the capital required to make this shift, and many more saw it as undesirable because it burdened the land with debt, subjected the enterprise to the vagaries of the market, and enmeshed relations between parents and children and among neighbors in a market calculus. Most families expanded their operations primarily to facilitate intergenerational transmission, enabling a son or daughter and his or her family to work alongside aging parents. The chronic instability of commodity markets and growing inequity in the distribution of incomes and property placed a premium on strategies that relied on local resources and were based on sharing labor. Cooperation made economic sense to most Midwesterners. From the decline in commodity prices after World War I through the Great Depression, survival depended on "making do," but the cumulative depletion of resources that ensued led to fundamental changes. The decline in the demand for hired labor stimulated outmigration even more than urban employment opportunities had done previously. The rural Midwest lost many young people, especially women of marriageable age. World War II, which brought unprecedented profits to some farmers, accelerated rather than reversed these trends. Farmers with more capital withdrew from local networks that employed poorer ones. Women's separate enterprises, especially their barnyard poultry flocks, were replaced by large-scale capitalist enterprises—the forerunners of the automated chicken-and-egg factories that now plague our food supply.

How did these fundamental changes take place? How did farming families resist capitalist transformation so successfully for so long? What roles did women play in sustaining diversified family farms and the community networks on which they relied? Looking at the connections between the family and the circle of neighbors proved particularly illuminating. But first, the complex gender and intergenerational relations within farm families and rural communities had to be sorted out. Thinking through the gendered political economy of farm families had bedeviled rural women's historians, who were acutely aware that there was no direct correspondence between the value of women's work and their say in farm family decision making. Women's toil both indoors and out did not always translate into power; sometimes it amounted to exploitation, and more often it was taken for granted rather than recognized. The connections between work and power could cut both ways. Some husbands presumed that their wives would do at least half the farm work as well as all the housework, but treated them as if they knew nothing about the business. Others allowed their wives to limit their own labors to the garden, barnyard, and kitchen, but listened to their opinions on everything. Rural women's historians puzzled over why some women seemed able to control their own labor and have a voice in the enterprise, while others seemed oppressed by a double burden of toil indoors and out and were disregarded or remained silent when decisions were made.[6]

Neth's conceptual breakthrough was to use the concept of negotiation to illuminate the connections among labor, value, and power:

> Farm women could use the economic importance of their labor to negotiate greater equality within legal and economic systems that were essentially patriarchal. Recognition, however, did not come automatically because of the importance of women's work. Women's historians have often assumed a link between women's "productive" labor and the status of women in the household—that if women's work was crucial to the family economy, or if women earned incomes, or if women participated in higher-prestige male labor, this was a sign of a more egalitarian family structure. Other historians have argued that male control of land and the legal system led to authoritarian households where women had little influence or power. In fact, power relations in families, as in other systems of labor organization, developed through a process of negotiation. When families are the organizers of labor, personal ties become the basis of economic life. Farm women's work and their evaluation of that work are inexorably linked to the relations of family and kin. Although many factors shaped the division of labor on family farms, the status of women and children within the farm family economy often rested on the quality of their relationships with the men who controlled farm resources. Promoting mutuality was a strategy that encouraged farm survival and improved the status of dependents within farm families. By emphasizing work flexibility, shared responsibilities, and mutual interests, farm people limited the conflicts created by the patriarchal structure of the family and agriculture and created strategies for the survival of family farms.[7]

The family labor system, as Neth explained it, had both hierarchical and egalitarian elements. Interdependence was a fundamental fact of farm life. Men controlled the capital resources, chiefly land, machinery, and loans. Most Midwestern families divided labor along lines of sex, although not all did so in the same way. Yet practices of cooperation in daily chores, especially milking, and in labor-intensive tasks, especially at harvest time, blurred gender distinctions and generated shared interests and common understandings. Similarly, commitment to the intergenerational transmission of the farm created habits of cooperation between parents and children, as well as among siblings.

Historians and sociologists who have examined these factors carefully have shown that variations arose primarily from differences in the farming family's class position, the commodities produced, and the composition of the family itself. More highly capitalized, less labor-intensive farms had a clearer gender division of labor. So did farms specializing in wheat, or corn

Figure 6. "Threshing wheat at the Beaujon farm, Endicott," New York, August 1945. Mrs. Beaujon (wearing overalls and kerchief) and neighbor woman (wearing hat) carry grain sacks from the threshing machine, which is tended by two men, to Evelyn Beaujon (wearing skirt), who loads them onto the truck. The Beaujons' farm was near the river along Campville Road (now Route 17C). *(Photograph by Charlotte Brooks, Standard Oil (New Jersey) collection, University of Louisville Photographic Archives. [SONJ 27842])*

and hogs. In contrast, dairy farms required all hands to work in the barn together twice a day and toil in the hayfields for a month every summer. Families with more daughters than sons, or with older daughters and younger sons, often treated at least one girl as if she were a boy; size and strength mattered more than sex. Flexibility was crucial. All labor was valuable, whether it was income-producing, subsistence-producing, or cash-saving. Income-producing operations were granted the highest priority by women as well as men. But women might be integrated into the primary operation, or they might conduct and control their own small-scale, market-oriented enterprises. And men as well as women were aware that subsistence production allowed the profits of the main operations to be reinvested in the enterprise.[8]

Reciprocity extended from kinship to neighboring, integrating work with sociability. Visiting, changing work, trading produce, engaging in mutual aid, and celebrating major life passages all generated a resilient network of interdependence among neighboring farm families. Gender-based and age-related networks linked individuals to their counterparts throughout the locality, complementing family-to-family relationships. The scale of work sharing was gendered; men combined forces periodically for the most labor-intensive tasks, while women more frequently mixed work with sociability while doing small-scale, portable hand labor, such as sewing. Forms of accounting also differed; men kept track of the relative contributions of and benefits to each party, while women more often assisted one another in keeping with norms of general reciprocity, for example, in caring for the sick. The tasks customarily assigned to men were subject to commodification and mechanization more rapidly than those allotted to women. At the same time, women's household work was more susceptible to being taken for granted as a "natural" activity without distinct economic value.[9]

Until World War II, norms and practices of interdependence functioned as an essential safety net for all family farmers and, for most, provided the basis for subsistence and the reproduction of both family and farm. This set of relationships served as a viable alternative to capitalist agribusiness, although it was capable of absorbing modern techniques such as mechanization and the application of chemical fertilizers. Equally important, it provided the foundation for farmers' organizing as the values they held dear and their vision of political economy based on landowning families were threatened by the forces of large-scale change. Radical farm organizations mobilized existing community networks to challenge the power of agribusiness. The consequences of their defeat are apparent across the landscape of rural America and in the vital role of off-farm labor for family survival.

The chapters in Part III analyze the division of labor and relations of power in farm families and their interconnections with the two main changes in the economy of the Nanticoke Valley during the early twentieth century: the increasing scale and degree of specialization of commercial farms, on the one hand; and the trend toward combining off-farm wage labor with small-scale farming, on the other. Taken together, these two shifts generated a new form of class stratification in rural society. To some degree, this divergence corresponded with ethnicity; many immigrant families who moved to run-down or abandoned farms kept some family members working in the factory while others labored on the land to build up the enterprise. But many of their impoverished native-born neighbors on hill farms also found it necessary to send family members to work in the city. The gender division of labor on these small-scale farms varied markedly: sometimes women and men worked together; sometimes the men did most of the farm labor while the women held off-farm jobs; and sometimes the women carried on the farm while the men worked in the city. Gender relations on large-scale farms that were more fully integrated into the capitalist market varied as well. The gender division of labor tended to be more clearly defined, as hired men substituted for wives and daughters in the fields and machinery replaced them in the barn. Yet women continued to exercise considerable power in farm family decision making. In many of these families, women worked alongside men when their children were young and stepped aside to make a place for their children as the next generation matured. As they aged, they still wielded significant power as business managers, as mediators of relationships between their husband and their children, and in the intergenerational transfer of the enterprise. Finally, until World War II women produced much of the family's living, which not only guaranteed an adequate subsistence but also enabled the returns from the main farm operations to be reinvested in the enterprise. The exigencies of the post–World War I decline in farm prices and the Great Depression meant that holding onto the land and accumulating capital depended on women's cash-saving as well as income-producing labor.

In the chapters that follow, I analyze these dynamics through extended personal narratives in which individual women articulated and interpreted their own experiences of work.

"I Just Never Felt That I Ought to Take from the Farm"

Putting the barn before the house seemed like simple common sense to women who founded farms with their husbands, whether they were newcomers like Josie Sulich Kuzma or had grown up in the community. Even a woman who did not routinely milk cows and do barn chores alongside her husband placed a higher priority on farm work than on housework and favored spending money on what the farm needed over raising their standard of living. Elizabeth Wheaton Graves, who was from a native-born family and married a man with even deeper roots in the Nanticoke Valley, founded a farm in partnership with her husband, Kenneth, during the depths of the Depression. Her narrative articulates the inextricable connection between a woman's marital partnership and her partnership in the enterprise.

Elizabeth's commitment to the farm is all the more remarkable because farming was Kenneth's dream, not Elizabeth's. While she had grown up in the village and held an office job in the city, he had spent his childhood on his grandmother's farm and worked for several farmers, including the progressive Edgar L. Vincent. Although he was a skilled tradesman at IBM when they married, "he had always thought he'd like to have a farm." He had been a live-in hired hand for a previous owner of the farm that Elizabeth and Kenneth bought "at the courthouse steps" for $7,500 in 1932, and "he loved this place."

Elizabeth's closest relatives and friends were opposed to the move. Not only were they were convinced that the Graveses would not be able to earn enough to pay for the land, but they worried about how much labor it would require. "I remember my aunt and my cousin saying, 'Oh, don't go on a farm, don't go on a farm, they work too hard, you'll work yourself to death,' and so on." The elderly neighbor in Maine village whom Elizabeth called "Grandma" expressed the same misgivings. At that time, Elizabeth explained, "they didn't have the equipment to work with, and women went out and worked in the fields and so on, you know. And she said, 'Oh, you don't have any idea what you'll have to do.'" Despite this well-meaning advice, "I said, this is what my husband wants to do; if this is what he wants to do, this is what will make him happy, this is what I'm going to do. So we came up here, bought the farm with nothing." They took a mortgage from friends of the man who had owned the farm when Kenneth had worked there, paying them in person every six months.

Before they could move in, they had to clean out the house; chickens had been kept in the back kitchen, and garden produce was rotting on the floors upstairs. It had no modern conveniences; for years they made do with a pump in the kitchen and a copper kettle to heat water on the stove. The land was easier to reclaim than the house, although paint had been dumped out back. The place had good bottomland along the creek for growing corn and other feed grains, fine pastures that extended up the gentle hills, and an ample woodlot, as well as an orchard near the horse barn. They always had a big garden, and Elizabeth canned vegetables and fruits for winter. Kenneth did most of the gardening and helped with the milking while the hired hands focused on the fieldwork and barn chores. The Graveses produced as much of their living as they could and devoted their income to buying cows.

Elizabeth told a series of stories about her misadventures on the farm during those first years that turned on her ineffectual efforts to help with the work. "After I came up here, I didn't know anything, really, about farming, and I wanted to learn to milk." Her father had kept a milk cow when she was a girl, but "I never learned to milk. My father didn't even like me to go to the barn. Girls didn't do things like that." "My father didn't even think I should be in the garden. He just was very old-fashioned about girls working outdoors. They didn't go barefoot, either! I loved to go barefoot out in the grass," she laughed. "Girls didn't do that. The boys could do those things, but girls didn't." Elizabeth was eager to work on the farm. "So I went down to the barn, and I was going to learn to milk. Well, I was afraid I was going to hurt the cow," she laughed. "And the [hired] men would laugh at me; they stood there laughing until I got disgusted. So I never did learn to milk very much. Later, I used to put the milkers on once in a while. But I used to drive the horses on the hay fork, and then run into the house and get the meals. And I learned to set up oats. I learned to drive the tractor. I guess I did everything but milking." The cattle were the responsibility of the hired men and her husband, who went to the barn every day before and after work. "To be honest, I just didn't have the time to do it anyway. We always had hired hands; I just had all I could do to cook their meals and wash their clothes and keep the house....And he [Kenneth] didn't want me to come down there and milk cows; he said that's what he was paying the men for." Elizabeth's desire to learn to milk was quashed by the humiliation she encountered in the barn.

"The second thing is, I didn't learn to split wood. I didn't want to learn to split wood. When we came here, there was just one big furnace downstairs, and we had a little stove in the kitchen. But that furnace would take wood this size"—she extended both her arms—"about all you could lift to put in. And I never learned to split wood." "Grandma had said, 'Don't learn to do those things, and then you won't have to do them.'" Ignorance of a skill was a wise strategy for avoiding taking on additional responsibilities; the older woman had warned that if a woman let on that she knows how to do something, she'd be expected to do it all the time.

Elizabeth tried to help out when emergencies arose during haying. "I remember one time, [when] our boys were about six and four," the hired hands were haying for a neighboring farmer who had helped them mow earlier in the day. "It looked like we were going to have a storm, and we had a lot of hay down on the north side of the house here. I said to the boys, I think we'd better see if we can get that hay in...and surprise Daddy. Well, we surprised Daddy all right, because we got the hay loader all tangled up. I think it broke some of it. We didn't get much of the hay in," she said, laughing. "But I tried to be so helpful." Another time the hired man who was doing the chores "came up to the house and said that one of the cows didn't come; 'she's way down in the pasture with her little calf.' So I...took him in the car, because you could put the calf in—we always used our car for everything—and we went down to the pasture. He said, 'I think it's kind of soft here, maybe you'd better be careful before you drive here.' I said, 'Oh, I'm sure this is all right'...and about then I went kerplunk! right in a mud hole. The hired man had to come up and get two of the men away from haying to come down. I was really too helpful, you know!" She laughed again. "I tried to help, but I didn't know enough to do it. Of course, I never heard the last of that, being so helpful getting the hay in. But there were a lot of things that I didn't know."

Learning to make butter was vitally important because the Graveses sold butter until they had enough cows and had improved the barn enough to sell fluid milk. Elizabeth recounted her initial effort at churning. "I'd never seen anybody make butter. Well, my uncle worked in the creamery down in Maine village, and my aunt knew how to make butter. She had a little glass churn with a little wooden handle....She let me take this little churn, and she said, 'Well, you set the milk in pans, and when the cream gets to the surface, you skim it.'...So I thought I was going to make a lot of butter. I put cream in this little jar and I worked and worked and worked at it." But no butter appeared. "Well, I thought the butter was going to come in a pound print, you know!" Again Elizabeth laughed at herself. "So I loaded the kids and butter churn in the car and I went down—I was almost in tears—and said, 'Auntie, I can't make butter, this doesn't make butter. I did just what you told me.' She laughed and she said, 'My dear, you've got butter.' She said, 'You've got to gather it, and pour the buttermilk off, and wash it.'...So that was my butter-making experience. I made lots of butter after that." When they had more milk, they bought a separator and a barrel churn with a hand crank. Elizabeth and Kenneth "took turns" separating the milk, and she did most of the churning. She worked the butter in a big wooden bowl with a paddle. For sale, she didn't mold it into one-pound prints, but put it in butter jars. The butter receipts made the single most important contribution to the farm income.

"After we got a dairy, and we couldn't send our milk, we didn't have the barn approved, so I made butter and I sold buttermilk down to the hospital in Endicott for a long time....And we had chickens. So I was a butter-and-egg lady....We sold eggs for years. Later, during the war, we sold hatching eggs, and I had to weigh them....They used to bring the crates up here, big loads of crates....I used to put them in the car and take them to Union; from the depot there, they shipped them to New York." Normally people came to the farm to buy eggs, as well as butter and buttermilk. Once the Graveses had more than two dozen cows and a water-cooled holding tank, they could sell fluid milk to a processor in Binghamton. Kenneth thought that they had made that transition by the early 1940s. Eventually he and the hired hands were milking between thirty and forty cows, and after they bought milking machines, Elizabeth helped out occasionally.

Elizabeth recalled cooking for the large crews of men who came at harvest time as her most burdensome task.

I remember Mr. Wakefield used to have a thrashing machine and go around and do the thrashing for the other farmers. He came over here and this one day I had twenty-two men for dinner. Well, I hadn't ever done anything like this for so many people, you know....I baked pies for a couple of days. And I put wash benches outside for them to wash in, since we didn't have a sink in the back room. That was my life, from one job to another. Well, not having daughters, or sisters or anybody, I did it all myself. This one time Mr. Wakefield went home and told his wife, "You know that young woman over there, she's getting meals for all these men," he says, "I think you ought to go on over and help her." Of course, Mrs. Wakefield did—she was a wonderful person—and she brought both her girls. The next day she came over with a great big coffee pot and said, "Elizabeth, I've come over to help you." I said, "Why, Grace, how nice of you, but I'm doing all right." "No, you're not; George said it was just terrible, that young woman over there doing all that work for all those men, and so he says I think you'd better take the girls and go over and help her." So they came over and helped me! Of course, the men worked back and forth a great deal.

One of Elizabeth and Kenneth's favorite inside jokes turns on the priority they placed on acquiring cattle rather than accumulating other forms of capital. "People said, 'Your husband works at IBM, you must have a lot of IBM stock.' I always said, all our stock is four-legged; everything went into that barn." Expenditures for the barn came before things for the house. "There wasn't much question. We had to have things for the barn. That was our livelihood. I mean, we were trying to pay for the farm. We went..." Here Elizabeth started to say that they "went without," but stopped, corrected herself, and continued: "I was never really deprived of a great many things, but we didn't modernize our kitchen for many years." With Kenneth as a silent partner in our interview, she explained, "If I wanted anything, I'd say to Kenneth, I wish I had...'All right, you get it, then.' But this is the way we did it. He'd always say, 'You get what you want.' And sometimes for an anniversary, he'd buy a silo or a manure-spreader!" She laughed, and he joined in. "But I was always perfectly willing for him to do that." Her husband trusted her not to spend money unnecessarily or to spend more than they could afford. "We put so much, all through the years, into the farm, because it was a necessity to begin with, and I have never felt deprived. A lot of my friends have things that I never felt that I...well, I just never felt that I ought to take from the farm."

The way that she purchased her washing machine speaks for itself. They had two children in diapers, Kenneth's grandmother was living with them, and Elizabeth was doing laundry for her elderly relatives. So "I took in washings, to buy a washing machine, when we first came up here. I got the washing machine, but that's the way I had to pay for it." She purchased the machine on the installment plan and used the money she earned doing laundry to make the payments. "One time I was doing five washings besides my own," including linens for a restaurant in the village. The machine "paid for itself"—at least, if you don't count the cost of Elizabeth's own labor.

Although Elizabeth became a farmer because that was what her husband wanted to do, she spoke for many rural women when she rejected the term farm wife, declaring emphatically, "I'm not married to the farm!" She understood why home economists had searched for another term to substitute for the vernacular housewife; women were not married to their houses either. But she regarded the term homemaker as a way of trying to dress up the ordinary tasks that urban as well as rural women performed. "It's not the work that makes a home anyway," she contended. In asserting not only that she was a farmer but also that men as well as women were, first and foremost, members of farming families, Elizabeth Wheaton Graves, like many Nanticoke Valley women, articulated and upheld an alternative set of ideas and practices about gender relations to those that predominated in American society during the early twentieth century.

"If You Think a Farm Woman Doesn't Milk, You're Crazy!"

Frances Sheldon Howard, like Elizabeth Wheaton Graves, founded a farm with her husband. She was born 1901 and lived in Maine for much of her childhood. In 1925, she and her husband bought a farm on Pollard Hill, where they lived until World War II. In her modest and unselfconscious narrative, she voiced the assumptions that rural families held and expressed astonishment that others might not share them.

The Sheldons were native-born, but whatever land the family had previously held had slipped out of their hands. Frances's father traveled around to work in the woods, and when she was between eight and ten, the family moved into a rented house on the Newark Valley Road west of Maine village. Her father continued to work as "an ordinary laborer," although it was difficult for landless families to make a living in the country. Her parents bought land on the old Newark Valley Road after she had left home.

At sixteen Frances went to work at Endicott Johnson "like everybody else did around Maine at that time. Because there was nothing around here, see." She and her sister roomed in Endicott at first; after the bus service started, she lived at home and paid her parents for her board. She often bought things the family needed as well. Frances "didn't feel particularly different from other people" in the city; "everybody seemed to be on the same social level, and nobody looked down on people from the country." After all, the majority of workers had come from rural areas.

She married Alfred Howard in 1920, when she was nineteen and he was twenty-one. He was working at what became IBM. "But he didn't like inside work, so he decided that he wanted to get out and do farm work." Although Frances "hadn't planned on farming," she "didn't mind." "It was he that made the decision to, what he wanted to do in life; why shouldn't you go along with it?" "When you get married, whatever livelihood your husband—he's going to be the breadwinner—feels he wants to do...that was the choice that he made, see. So I guess maybe it was the right one, because we didn't make out too bad." Frances evaluated this decision not only in retrospect but also in relation to what she felt was a fundamental principle: "When you're married, you know, you don't work against each other, you just work with each other."

The 180-acre farm they bought had some valley land, 10 acres of meadow, and ample pasture on the hillside. They started out with about

20 cows and enlarged the herd until they were milking about 25. They milked year-round, "so that you had a steady income," and took the milk to the creamery in Maine village. They cultivated corn for silage and oats for feed, although they had to buy supplements from the Grange League Federation (GLF) cooperative. Their third child was born the spring they bought the farm; later, they had a fourth daughter.

Frances said emphatically that she always milked: "If you think a farm woman doesn't milk, you're crazy!" I explained that I asked whether she went to the barn because I had observed remarkable variations in how different families organized their work. She responded by offering her own generalization: "Every farmer's wife that I knew, that we were acquainted with at the time, they all went to the barn and milked and helped with chores, always. Good Lord, yes....I always used to put the cows in and start milking before anybody else came to the barn, you know what I mean? Because you were able to do it. Gee whiz, that meant a lot less work for somebody else; they could work in the fields...." Frances "didn't work in the fields very often," although she knew women who did. Polish women on the adjacent hill farms generally worked outdoors; "They were strong, and that was their nature." Frances worked alongside her husband in the barn twice a day, and their daughters did chores before and after school. They seldom hired help. As she showed me a photograph of their calves, she commented admiringly, "Here's some of the white faces we raised; ain't that pretty?"

When they bought the farm, it was quite run down. "After we bought the place, Mr. Paine told us we couldn't raise an umbrella on it. But my husband bought a lot of lime and fertilizer, and we had some good crops there, boy, I'll tell you. But it was because the other people didn't do anything to feed the soil; if you feed the soil you're bound to get a harvest out of it." They bought a tractor around 1930; she never learned to drive it because "there was always somebody else around to do it." She and her husband gardened together, especially after chores and supper. They also kept chickens. In the early 1940s, when the price that the GLF paid for eggs had fallen, Frances started an egg route of her own. "I got market price for them, what our local market charged. So I made a little money off of those." She drove the car on her route every Friday.

At first their house had no running water; it had a pump outside on the porch, and "it was awful cold to go out there and pump water in the wintertime." After the electric line came through around 1929, they had a well drilled and had water pumped into both the barn and the house. Frances still took the milking implements up to the house to wash, however, because they had to be sterilized in hot water. They bought milking machines around 1935, about the same time they installed a coal furnace.

Frances said that improvements to the barn came before those for the house—or at least along with them, as the running water did. To her way of thinking, money spent for farm improvements was invested, but money spent on the house was merely spent. "If it was an essential thing, you wouldn't frown on it, if it meant your livelihood. You shouldn't feel otherwise. You had to" put the barn ahead of the house, "because that's where your income was coming from, your cows and things in that category, and they had to be taken care of whether anything else was or not, you know. That was your mainstay." Frances knew what the farm needed. Her husband was willing to spend money on modernizing the house: "When we could afford" to buy things, "we had them. Of course, a lot of times you might want things but you couldn't afford them, so you'd wait till you could." Frances said that she and her husband always talked over big decisions together.

Frances's idea of marriage meant that husband and wife pulled together rather than apart. Their partnership enabled them to manage heavy burdens, acquiring land and building up a commercial dairy herd. Although Frances used conventional language in describing the man as "the breadwinner," once her husband decided to farm she became a full partner in the undertaking. She worked alongside him in the barn and regarded it as equally natural that they made decisions jointly.

"Woman's Proper Place Was Wherever She Was Needed"

Caroline Harris Bevier Stevens, born in 1914, worked on her parents' and relatives' farms during her childhood. The Harrises were landless, so they farmed on shares at first. In 1918, they moved into Endicott, where both her parents held paid jobs. Eleven years later, they traded their house for an 80-acre farm on Ashley Road in the town of Maine. "My Dad wanted to get back into farming; he wanted a place that he could call his own." Her father kept his job at Union Forging, however, so her mother did most of the barn chores and much of farm work. By that time, Caroline had graduated from the eighth grade and was doing housework and childcare for pay. After they moved to the farm, she stayed home to help. Caroline proudly recalled how hard she and her mother worked. In the country, "woman's proper place was wherever she was needed to be, and that was twenty-four hours a day." When it came to farm work, "It didn't make any difference if you was a girl."

Both Caroline and her mother worked in the barn and the fields, as well as in the garden. "Everything was raised for winter; very little was bought." The children did the weeding, "sometimes all day long when there wasn't other things to do." Caroline's mother "used to crab about my not working in the garden," which she did not enjoy. "But I worked in the garden when I had to, and I canned a lot of stuff." They also cultivated potatoes: "Oh, boy, did I dig potatoes. We planted and dug potatoes all by hand." During the Depression, the family always had plenty to eat, and her father would offer a meal to any tramp who came by. The farm they had rented had a large enough dairy herd to take milk to the creamery, but in Maine they never had more than five cows. "We didn't have a milking machine and all like that. My mother could milk two cows to my father's one, and you did it by hand." Her mother used a mechanical separator and a barrel churn to make butter. Her father sold their butter and cottage cheese, eggs and chickens, and potatoes to customers in Endicott, along with stove wood that he logged, split, and bundled in cords.

All the farm work was done by hand because they could not afford any machinery. "We took turns" mowing, tedding, and bunching the hay and pitching it up onto the wagon and into the haymow. Caroline drove the horses to pull the mowing machine: "You took your turn; whoever was available did it." She also worked in the fields during the harvest. "I didn't cut oats or corn. However, I did have to take the oats and get a bundle of them and make a tie for them and tie it, and then stand it up." Her parents took their corn and oats to Pitcher's Mill to be ground into feed. Caroline also worked with her father in the woods.

The small scale and diversified character of their farm meant that everyone's labor counted. Significantly, she was not the oldest child at home at the time; she had older and younger brothers who worked outdoors just as hard as she did, although they did not work in the house. Caroline helped her mother with the housework when she had to. "I cooked and baked, and scrubbed the wood floors on my hands and knees with home-made soap." The family had no electricity, no inside plumbing, no running water, and no telephone. They "didn't even have a pump in the kitchen" at first, but carried water into the house from the pump out behind the summer kitchen. When I asked Caroline if her mother would have liked to have household conveniences, she responded, "I don't think that she really...I think that was her way of life. I don't know, she probably would have" appreciated them, "but they were out of reach." Her parents still had no electricity when her mother died in 1950. Her mother did have a sewing machine and made the children's clothing. Caroline recalled wearing dresses made out of flour and feed sacks. She did not regard them as a sign of poverty; she thought they were pretty and assured me that after repeated washings they were not too scratchy.

Caroline preferred working outdoors to doing housework. The physical freedom it afforded her was one reason; the other was her parents' differing temperaments. "I would rather work with my father than my mother. My father wasn't as fussy as my mother." Her mother had high standards for housework. "I never quite satisfied her, because...well, that's the way she did it, and there was no other way to do it." At times she would rather do something herself than try to teach her daughter how to do it properly—"as long as it wasn't washing dishes; she didn't like to wash them any better than I did." Caroline's father "wasn't as persnickety" as her mother. "But he was very fussy about feeding the animals. You did that at the time it was supposed to be done. And no matter how deep the snow was, if the animals couldn't go out to go over to the creek to drink, you carried water to them, and you carried till they quit drinking."

When I asked who had the most authority in her family, Caroline replied immediately, "Why, my mother was boss. She was boss; it was easier for Dad to go along with her decisions." Her parents had agreed to leave the first farm because "they couldn't make enough money farming on shares." When it came to moving to the farm in Maine, "My mother was all in favor of getting on a farm. It was getting pretty bad down in Endicott at that time....worked part time, part time you didn't. So I guess that both of them decided that together, to get out somewhere where they could have a few acres and do something on it." As far as day-to-day decisions were concerned, however, "I think that she was the one whom Dad went along with."

While her parents were still in Endicott, Caroline went to live and work on her relatives' farms, first with an aunt and uncle whose children were very young and then with another aunt and uncle who were childless. "They needed the help," she explained, so when she was thirteen she replaced her older brother on their farm. She learned to milk cows and tend chickens, as well as helping with the housework. Later on she worked for wages for the Martin family, who ran a retail dairy. She took care of their son "and did housework, because" Mildred Martin "worked outdoors most of the time." Although Caroline married a man who did not farm, "I worked outdoors then, too, because his brother had the farm right next door, and I used to help with the haying and so on." She also earned money by helping other neighboring families during haying and harvest. She made sure I understood that she "didn't cook for the men" but toiled alongside them in the fields and meadows.

Caroline Harris Bevier Stevens's experiences and perspective are typical of women who grew up in the Nanticoke Valley. The exigencies of small-scale, diversified farming placed a premium on labor power, and workers' sex mattered much less than their skill, strength, and reliability. Dairy cows had to be milked, fed, and watered twice a day, creating a kind of chronic emergency that took precedence over everything else. Almost everyone who grew up on a farm learned to milk and expected to do their share of barn chores. The daily sharing of labor that dairying required generated habits of cooperation that extended to other forms of work as well. Haying was a seasonal emergency, when all hands turned out to cut, turn, and get in the hay before it could be rained on. Indeed, women performed some of the most laborious tasks outdoors because those demanded extra hands. Even though machinery accelerated some tasks, it also created more manual work; for example, women, children, and older men bound grain or raked and piled hay that had been cut by horse-drawn reapers or mowers. The allocation of work was not defined by sex or age; rather, family members took turns at specific jobs. Tending the poultry flock and the garden, as well as cutting firewood for sale, were tasks that parents and children shared.

Other chores, especially housework, were fitted in around these more important tasks. Like most women in the Nanticoke Valley, Caroline regarded
the popular notion that a woman's place was in the home as inconceivable.

"They Weren't Large Girls, but They Loved to Work Outdoors"

Violet Burton Canaday's family lived in Binghamton, but they spent lots of
time on her grandparents' farm in Maine because her grandmother was not
well. When she was a girl, Violet "used to love to be outdoors," so her memories of the farm are particularly vivid. In recounting her mother's memories
of growing up on the farm and going to school in town, however, Violet's
narrative was marked by contradictions that registered the gap between
rural practices and norms and urban, middle-class ideas about womanhood.

Violet's mother did so much farm work because the Ashburys had a relatively large-scale dairy operation and their four oldest children were daughters. Two elder sons had died; two more sons were born later. "They weren't
large girls," Violet said, "but they loved to work outdoors." Their labor was
especially important when they were adolescents because their mother was
not strong and their brothers were still small. The family had forty cows.
"The girls each milked ten cows night and morning." Their father "milked
the ten hardest. And then he always insisted that they have an education, so
they would get up and milk the cows and then drive the milk to the creamery and go on to high school" in town. Three of the daughters became teachers, and Violet's mother worked in an office. Neither of the sons wanted to
go into farming. "Farming was hard labor. . . . You don't get to go away when
you're farming. With cattle and everything, you've got to be there and see to
them. Even on Sunday."

In later years, her grandfather bought a tractor and a hay loader. "He
always had the latest equipment, as much as he could" afford. He "put in a
milking machine that ran on batteries." They also had "a washing machine
that ran with a gasoline engine" outdoors. The electric line "didn't reach the
farm" until the mid-1940s. They put running water in the barn, and the milk
house had a deep well for cooling the milk. "That's the kind of people they
were; if there was any way they could have anything, they had it." Still, the
house had no inside plumbing or central heating.

The allocation of tasks within the family was very flexible: "Anybody
that came along" did whatever needed to be done; "there wasn't anything
set." When they churned butter, "anybody that felt up to it" would do it.
"My grandfather would do it, and my grandmother would do it; anybody
that could do it, did do it." Violet's grandmother kept chickens, but she did
little other work outdoors because of her chronic ill health. Although it
turned out that she had pernicious anemia, she was initially diagnosed with
a heart condition and advised to avoid strenuous physical labor. With so

many children, she had enough to do in the house anyway. "Because she sent them to school in white stockings, oh yes....They were educated people; they weren't people who didn't care how they were, who just stayed on the farm and that was it." Her grandmother's mother had been a teacher. Her grandmother had strong political views that differed from her husband's; "one was Republican and the other Democrat, so they cancelled out each other's votes!" She kept the house immaculate and often had it painted and papered, although she had help for spring and fall cleaning. "My grandfather was very good to my grandmother. He would go down cellar" to get canned goods for her, for example. Violet recalled that her grandmother "always had money" to spend on things for the children or the house. Her grandparents talked over decisions about the farm. "She was quite liberated for her time; she wasn't put down at all."

As a girl, Violet's mother not only helped rake hay but also plowed the fields. Once "she plowed a twenty-acre field so" her younger brother "could go to a baseball game!" The oldest daughter helped their mother in the house, but the others preferred to work outside. Their mother "didn't mind." When I asked if she did not insist on their being ladylike, Violet replied, "I guess you couldn't be a lady even if you wanted to!" Yet, she recalled, "my mother said that she never went to school without having a dirty neck" because she drove the horses to school and left the milk cans at the creamery on the way.

The conflict between urban and rural ideals of womanhood is evident in this predicament. A young woman who milked every morning could wash up and change her dress before setting off to school in town, but the reins would leave a residue on her hands, which would smudge her neck when she brushed back her hair. This sign betrayed her as a farm woman, no matter how cultivated her mind and refined her manners. Another aspect of Violet's narrative shows a conflict between urban and rural ideas about women's health. In her stories about the farm, Violet's mother always said that working outdoors made the sisters strong and healthy, developed their self-confidence, and offered them freedom from the restraint that was imposed on city girls. As adults, however, they were not so robust. The city doctor offered a common explanation for the sisters' difficulties; the heavy physical labor they had done in their youth had overtaxed their bodies. To understand how Violet could believe these two contradictory things simultaneously, we must understand her personal experiences as well.

Violet and her mother both had problems with childbearing. Violet had rheumatic fever in 1942, which made pregnancy dangerous, and her first baby died because of placenta previa. She identified her own "female troubles" with her mother's reproductive history. Her mother had four children, but two of them were stillborn. "My mother was never a large woman. And I think that the doctors always said that the hard work that the girls did was harmful to them. They weren't big, strong girls. Of course, nobody realized this; they liked to do the work, it isn't as though they didn't like to do it. For they'd rather go out and do the work outside than work in the house." "The thing was that they said some of my mother's troubles was caused by it....That's what they laid it to: lifting heavy milk cans and doing heavy work."

The discrepancy between her mother's recollections and her acceptance of the doctors' interpretation was so striking that I questioned Violet closely. She explained that "one aunt died when she was twenty-seven of TB, and the other died of cancer when she was thirty-four or thirty-five. The other died as a result of a fire." I pointed out that none of these conditions was caused by overwork. But Violet was aware that her aunts had also suffered from debilitating pain, a prolapsed uterus, and multiple miscarriages. Her mother became convinced that these problems were the result of "having been forced to work too hard when they were young"—language that suggests that women would never have done these tasks voluntarily and implicitly blames their men for expecting them to. Lifting milk cans was too much for women's bodies, her mother had declared, and might cause permanent damage to their reproductive system. This opinion came from a male physician in the city whom Violet's mother had consulted in adulthood. She gave this pronouncement credence and had communicated it to her sisters as well as her daughter. At the same time, she continued to tell stories about the pleasures of working outdoors on the Ashbury farm, contrasting her current poor health with the vitality she had enjoyed in her youth.

These discrepancies reveal a contradiction that Violet's mother experienced throughout her life, not simply a change in her ideas over time. The intersection of rural and urban belief systems occurred every day as she took milk cans to the creamery on her way to high school. Violet's mother tried to fulfill the norms both of the country and the city. But she was discomfited by the cultural distance between them, which seemed indelibly inscribed on her body as she traveled from one place to the other. Similarly, a woman who gloried in the strength she developed while working outdoors as a young woman blamed the reproductive problems from which she suffered later in life on rural residents' ignorance of how heavy labor might damage women's bodies. These two convictions coexisted in Violet's consciousness, just as they had coexisted in her mother's remembered experience.

6

※

Intergenerational and
Marital Partnerships

What shaped the gender division of labor in farming families? How
did their work affect and reflect the everyday interactions between
husbands and wives? Looking closely at the working relationships
between women and men on a range of family farms reveals several clear
patterns. First and foremost, the flexibility of labor was crucial, especially in
labor-intensive operations such as dairying. Few strictures about "ladylike"
behavior constrained rural women when it came to doing what needed to be
done. Second, women's work was more central to smaller-scale, more diversified enterprises than to larger-scale, more specialized ones. When another
full-time worker became necessary as farmers expanded their operations,
a hired hand might replace a woman who had other things to do. All farm
women conducted subsistence operations, and many had independent, market-oriented "sidelines" as well. How much farm labor they performed depended on their own preferences and on the makeup of the farm household.

Because gender distinctions were almost always secondary to the farm's
need for labor, the composition of the household affected the work assigned
to growing children. In families that had no sons, or that had several daughters who were older than any sons, at least one daughter worked outside
with her father rather than in the house with her mother. Intergenerational
relationships had an equally powerful influence. Mothers, like fathers, were
deeply concerned with farm succession. Inducing at least one child to remain on the farm and enabling him or her to take over the enterprise while
supporting the older generation was a key goal of those who had inherited their land. Over time, a woman who routinely went to the barn and

occasionally worked in the fields when her children were young might step
back to make room for the rising generation and eventually allow her son
and/or daughter and son-in-law to replace her as farm partner. The interactions of these factors created substantial variations in families' working
relationships over time.[1]

Stepping Back to Make Room for "Her Boys"

Gender relations in the McGregor family exemplify the shifts that occurred
as a family enlarged its farm operation and became an intergenerational
partnership. Clara Durfee McGregor, who had grown up on a small farm in
the Nanticoke Valley, worked closely with her husband, Venley, during the
decade before they had children. Married in 1901 when both were twenty-
one, they started out with a diversified farm on a relatively small plot of
land near Maine village. They had a market garden and dairy herd as well
as a poultry flock, and they peddled their produce in Endicott and nearby
towns. Then they specialized in poultry, raising their own hens, selling eggs
wholesale, and breeding chicks to supply women with barnyard flocks. The
McGregors were recognized as progressive farmers both regionally and na-
tionally. Venley was chosen a Master Farmer in 1934; the next year they
went to Washington, D.C., for the award ceremony at the White House.
Clara noted in her diary that she was "very proud and happy to be part of
it" when she and other wives of Master Farmers had dinner with Eleanor
Roosevelt at Cornell in February 1937. Venley emphasized the familial character of their enterprise when he was interviewed by the *American Agriculturist*: "Whatever success we have had looks small to us compared to the joy
of having our boys come back and go into business with us."[2]

When I interviewed two of the McGregors' sons and their wives about
Clara's role on the farm, her sons saw the situation rather differently than
her daughters-in-law did.[3] The openness with which they debated the matter was equally striking. The middle son, Dane, declared that having children "was the big want of her life" before she took in her orphaned niece
and then had three sons of her own.[4] Margaret, who had married Dane's
elder brother, Garth, said that "It was 'her men' that she talked about—her
three boys and the father—'the men,' 'my men,' you know." Ruth, Dane's
wife, agreed that Clara always spoke of what "the men" wanted, needed,
or would like; she "would do anything for her 'boys.'" When I asked them
to describe Clara's work, Dane began, "She helped some with the chores
and with the chickens, but primarily she was a housewife and mother." I
then inquired what specific farm tasks she performed because I knew that
some women who worked alongside their husbands outdoors nearly every
day were described by the menfolk as "helping" them. He replied, "Well, I
remember her going out, when we had chickens on range, going out and
watering the chickens, feeding them, and so on, before we got big enough

Figure 7. McGregor family, Maine, New York, ca. 1934. Seated in center: Clara Durfee McGregor and Venley McGregor. Standing behind (left to right): Warren McGregor, Ruth Tibbets, Margaret Murphy McGregor, her husband Garth McGregor, Ruth Reese McGregor, her husband Dane McGregor. Seated in front: three McGregor grandchildren. Photographer unknown. *(Courtesy of the Nanticoke Valley Historical Society.)*

that the men were doing it. She worked with the baby chicks some, when the hatches came off, and helped clean up and so forth. As we got bigger, she didn't do any of the work outside." That is, when the farm expanded and the boys matured, she stopped working outdoors. Ruth then added tactfully, "I think your mother did work hard, worked right along with your father before any children came." Dane conceded that "from what I've heard, I'm sure she did." Later, when her sons were in high school, she again "helped out" more. Margaret concurred that Clara was integrally involved in the poultry operation before the children were born and when they were young. Later "she did a lot of the secretarial work.... She didn't keep the books, but when they'd have baby chick orders come in she would type the letters. She enjoyed that, and did a really good job of it."

Her sons perceived Clara as relatively uninvolved in the operation because she did not tend the incubators and brooders, as the men did, or go out to the egg room, where hired women candled, washed, and graded the eggs. They acknowledged that she helped care for the poultry when they were away, but this labor was invisible to them. In their minds, Clara was a model of domesticity. "She left the decision-making entirely up to my father," Dane stated; "she was very dependent upon him, always." Ruth agreed that Clara often deferred to her husband's opinion. Margaret also described Clara as "leaving the business decisions to Venley. But," she quickly added, "she always knew all the facets of the business."

Her daughters-in-law saw Clara's ability to leave the outdoor tasks to others as largely a result of the family's composition and class position, but to her sons it was part of their shared value system. The younger McGregor men regarded their father's considerate treatment of their mother as a sign of the "respect" that women deserved and of the men's own respectability. In their eyes, their father placed Clara's needs and comforts first. The daughters-in-law demurred from that note of condescension, seeing Clara as actively involved in the poultry operation before she turned these tasks over to her maturing sons, well informed about the enterprise, and able to make her opinions known, however unobtrusively.

Clara Durfee McGregor's 1937 diary underlines the accuracy of the younger women's perceptions.[5] Not only does it document her regular participation in preparing poultry for market and handling business correspondence, but it creates the impression that, without ever challenging or even seeming to question her husband, Clara ruled the roost. What she wanted never had to be directly stated but, nonetheless, prevailed. Much of Clara's power came from the authority she enjoyed as a mother and from her central role in the interactions among her husband and sons. It was Clara who mediated the tensions among the brothers and between them and their father that arose from his control over farm management and their uncertainty about their personal futures. She was involved in her sons' deliberations about whether to go to college and when to marry. Clara handled interpersonal relationships well, and in so doing she achieved her own ends without calling attention to them.

In 1937, all three sons were working with their parents on the farm. The youngest, Warren, was finishing high school. Dane and Ruth were living upstairs in an apartment, and Garth and Margaret were living in a house next door. The daughters-in-law ran their own households, although they often shared meals with Clara and Venley and entertained visitors together. Clara's diary, which covers January through June, lists her own work and the activities of all the other members of the family, as well as their interactions with other people. This matter-of-fact account contains few expressions of emotion or statements of opinion, but the patterns it records are revealing.

The McGregors' house was much more modern than those of their neighbors. They installed utilities and purchased labor-saving devices before anyone else in the vicinity, even when it meant using money from a mortgage on the land. Dane said, "We had running water and inside plumbing when I was five years old, back in 1919." They set up their own electric plant five years before the power line came through the valley. It not only powered the incubators and brooders but also carried current into the house. After Venley paid off the first mortgage on the farm, the business suffered a financial setback, so at the onset of the Depression he took out another mortgage, which was paid off by 1933. Although most of the money was invested in expanding the poultry operation, they also modernized the house. Ruth took up the story: "They had the inside of the house done over. I know that impressed me so much. They had sand-plastered walls put in, and hardwood floors put in, and new furniture, and rugs, and so forth. I thought that this was the most elegantly fixed place in the town of Maine. But to think that they had the money to do those things!" Ruth was struck by the contrast between her parents' house and the McGregors. "The night before I was married was the last night that I took a bath in a wash-tub," she said with a smile; "We didn't have inside plumbing." She was astounded that a farm family could enjoy the same amenities that city people had. Strikingly, she recalled that Venley "didn't much care about having beautiful things around him." The impetus for improving their home came from Clara, and Venley was happy to go along. Clara's 1937 diary records continuous small improvements to the house. Spring cleaning entailed buying and putting up new wallpaper, shades, and curtains. They purchased new linoleum in Endicott and bathroom fixtures in Binghamton. There were occasional vexations: "Men had left the stair door open, plaster dust all over varnished floor. So very aggravating." Both of the younger couples purchased new furniture and appliances during that year. Clara was particularly pleased when Ruth got a refrigerator: "That will be swell for both of us."

Ruth said that Clara "always had a hired girl to help her" in the house. Dane explained, "She wasn't rugged, and she had a hired girl to come clean and iron and such for many years." Clara's niece did not help much in the house, but a neighbor girl, Mildred Kenyon, used to come over to work after school. Mildred recalled that Mrs. McGregor was "very busy with the chickens and the children," so she helped do the housework and took the baby out in the carriage. Later on, according to Dane, "Lena Cornell helped in the house as well as working in the egg room." In 1937, Lena was there every day, another woman did the laundry and other heavy chores, and a third came for the spring and fall housecleaning. On occasion, Venley himself helped prepare and clean up after meals—a remarkable turnabout because in most farm families women helped men outdoors more often than men helped women in the house. Significantly, although Clara often referred to "the boys" collectively, only once did she refer to her two daughters-in-law as "the girls." In that instance, "Dad, the girls and I did the dishes" while "the boys" went out.

Clara's work load was limited in part in deference to her apparent frailty. "I don't ever remember her being strong and healthy," Dane said. This notion may have developed during the decade after her marriage. Her first pregnancy ended in a miscarriage when she was six months along, and she did not conceive again for many years. In her diary, Clara frequently mentions having a "hard headache," although they seldom lasted the whole day. Occasionally she said she "felt grippy" or blamed her liver, but sometimes she said simply that she felt "blue" or "nervous." Most often, Clara listed everything she had done in a day and declared categorically that it was "*too much.*" Her usual response to a sick headache was to stop working and rest, which was possible in large part because Lena could take over the necessary housework. When she felt "weak and miserable" after a bad night, she "lay on the davenport most of the day." Another day, after she "woke up with very severe headache, Dad brought my breakfast in bed." When Clara was cleaning the house before having the floors varnished, she "got so tired I could hardly navigate at all. Dad came in to help with supper, bless him." The most telling entry comes during spring cleaning. "I cleaned one more oblong of ceiling, so dizzy and tired I almost fainted, so went on a 'sit down strike' for a while." In using this term, Clara likened her own work stoppage to the mass strikes staged by automobile workers in Detroit, who were making headlines at the time. Still, she mopped the kitchen and got dinner. Then "Warren scrubbed ceiling in PM, did a swell job. Mrs. Newton and Dad did an hours work trying to improve it."

Clara remained actively involved in the poultry business and recorded doing some sort of farm work every few days. In January, she dealt with customers who came, called, or wrote to order chicks and broilers; noted when they got an unusually large number of eggs (2,400 on a single day); recorded how many coops full of baby chicks "Ruth helped Dane to put out"; typed business letters and "helped Dad" with the bookkeeping; co-ordinated the design, printing, and mailing of an advertising circular; and recorded when they "picked and dressed chickens" to get them ready to take to market, a messy task that involved plucking and singeing off the feathers and removing the intestines intact. In May and June, Clara frequently prepared chickens for sale, sometimes but not always with another family member. One day, "Didn't get started picking broilers till afternoon. Did 150 of them, which was some job. Had to leave hens till morning when I'll have to clean panty. Dad and W. candled eggs tonight, I went out with them awhile." Holding up the eggs to a bright light (formerly from a candle) was necessary to detect and eliminate those that were rotten. Another day, with Ruth's help, "we got all the 300 chickens picked." The next week, "They picked a few roosters while I was getting dinner, we picked 163 in PM with most of them to catch."

Although Clara occasionally accompanied Venley on business trips around the county, she usually coordinated everything at home. Sometimes the house felt like Grand Central Station: "Loads of phone calls and ways to turn all day. Got lunch for Dad and Dane at 11, dinner at 12, dinner for Dad and Dane at 2. They had hard trip to Little Meadows with chicks. Warren took chicks to DuBoises." Keeping track of things when the men were going off in different directions was an important responsibility, and it ensured that she was well informed about the business. Clara detailed the improvements they made in the poultry operation in 1937: installing a second incubator, putting heating pipes in the cellar where they often picked chickens, and renovating the old creamery building for farm use. She recorded the ongoing negotiations about the land that the town of Maine took when it widened the road, which involved a survey, appraisal, and agreement on compensation. She also noted when "Mr. Thomas here for signature on Federal loan on the place."

Clara's husband and sons were careful not to burden her unduly. Although bouts of intense farm labor were unavoidable, the family could afford to take some time for rest and leisure. For example, after "we picked 175 fowls, men all insisted I go when Dad delivered broilers late in PM to Progress, so I did. Everything worked just right. Dad took me to Dairy Kitchen for supper, a great treat." Of necessity, family members coordinated their activities. When "Dad and Dane" were remodeling a farm building, she "had to catch 70 broilers and 30 hens and pick them in PM. Had been warm, so we picked in barn but grew cold, I almost froze, doctored up so didn't catch cold....Warren helped get supper." Cooperation was typical of farming families, but in the McGregor household it went both ways more often than it did in many others.[6] The men sometimes helped Clara in the kitchen so she could help them in turn. For example, Venley "came in to help get dinner on, as he wanted Dane and Ruth and I to go with him after a cockerel."

Clara McGregor did not separate her own interests from those of the farm. Tellingly, on Venley's birthday "he bought a machine he had always wanted for my gift to him. He and Warren both delighted with it." The household was integral to the poultry operation; she often picked and drew dozens of chickens, packed hundreds of peeps, and put thousands of eggs in crates before they went off to market. These messy tasks preempted her time and took over the kitchen. After she pulled back from the outdoor labor to make a place for her sons, she remained deeply involved in the business. Her power was invisible but ubiquitous, expressed and confirmed in the reciprocal deference that characterized her interactions with her husband.

Women who lived on large-scale, specialized, capital-intensive farms were generally less involved in agricultural operations than those on smaller, diversified farms. The process of expansion often relegated women's work—and, sometimes, their opinions—to a marginal role in the family enterprise. What is most striking about the McGregors, however, is that Clara was never uninformed about farm operations or peripheral to decision making, even when she did less outdoor work than before. The McGregor family history attests to the power that a woman might exercise both as her husband's partner and as her sons' mother. Negotiating from a position of unquestioned respect, she enjoyed considerable deference from the men in the family and used it to limit her labor while remaining central to the enterprise. This family story illuminates the connections between the gender division of labor and patterns of family interaction across gender and generational lines, as well as the discrepancies between the work women actually did and how others perceived their role in the family enterprise.

"I Guess I've Done Most Everything"

Mildred Gale Martin, like Clara Durfee McGregor, helped build up a specialized commercial farm. In her experience, most women went to the barn and helped their husbands in the fields, as well as doing some of the marketing. Labor was divided by gender only in that most men did not do housework. Unlike Clara, Mildred always put the farm work before her housework, not only because it was more valuable but also because she preferred working outdoors.

Mildred was born in 1896 on a farm "way up the hill" in the town of Maine. Her parents never had more than about ten cows, so they made butter for sale. Her father separated the milk, churned, washed, and worked the butter, and packed it into five-pound jars that he sold in Binghamton. At the age of sixteen, Mildred married Thomas Martin, who lived on a neighboring farm. In 1921, they bought land on the valley floor and embarked on dairy farming. They started out with a long narrow strip of land along the road and eventually bought more land across the road toward the creek. With 150 acres of bottomland, they were able to raise hay, oats and buckwheat, and corn for silage, although they bought feed supplements through the GLF cooperative. Before they had enough cows to sell milk, they "used to take milk and buttermilk and butter and sell it" at the public market in Endicott. "We used to churn about 50 or 60 pounds of butter at once. And we raised all kinds of garden stuff and took that along with us, lots of potatoes. This whole field," she gestured out the window, "we used to have in strawberries. Then that whole yard," she gestured in the other direction, "was a chicken yard; we used to raise about 2,000 chickens a year." Mildred cleaned the henhouses, gathered the eggs, and washed and packed them in 30-dozen crates for shipment to New York City. She also killed, picked, and dressed chickens for sale. Both butter-making and chicken-picking were done in the kitchen, so it was often "an absolute mess."

Mildred milked cows and did barn chores with her husband. She quit milking only in 1944 when they bought milking machines, which "were awfully heavy for a woman to lift." She continued to wash the cows' udders before the milkers were put on and to carry the milk to the cooling tank, as well as distributing feed and cleaning the barn. In the mid-1920s, when they finally had enough cows, they took their milk to the local creamery. In 1931, after a friend suggested they could do better if they went into business for themselves, the Martins started their own retail dairy. The Meadow Creek Dairy delivered milk year-round. "For about twenty-one years we peddled milk. I drove the truck and had a boy deliver."

"I worked hard all my life," Mildred declared. "I plowed, and I ran the binder while my husband set the stuff up. I guess I've done most everything." She worked outdoors even when she was pregnant or had a baby to tend. When she was expecting her second child, "I loaded oats all that summer, even up to a week before he was born. I was awful big, but I guess I must have felt well and strong. I used to get up in the mow with the hay. I didn't mind going up, but I was afraid to get down, and I'd cry. I don't know what good it did, but I'd cry." Mildred was rarely daunted by the physical demands of farming and integrated the children into her routine as soon as possible. "My oldest boy, when I dug potatoes, I had to put him into a potato crate, because he couldn't sit up alone." When I asked whether she mowed and raked hay as well, she responded, "Did I rake hay! Get on the hay rake in the morning and never get off till noon and night. My boys had to be taken care of by my husband's brother; he was maybe eight or nine years old, and he used to wheel him around. My husband's mother had to work, too, out in the hayfield." The two families farmed separately, but hayed together. Her mother-in-law "had to take time off" to make a noon dinner for all of them. During thrashing and silo-filling, "you certainly had to put on a big dinner for a lot of men."

The house had basic utilities but few conveniences. When they bought the farm, the house already had running water. They put in a "pipeless furnace," but it didn't heat much of the house; water in their downstairs bedroom froze during the winter. To save money, they lived in four rooms downstairs and rented out the upstairs as a separate apartment. The electric line came up their road relatively early, and they replaced their hand-cranked washer with a power machine. They installed a bathroom in 1931. But "it was a long time before we had an electric stove or a refrigerator or anything." The barn always came first. "Well, I guess it has to," Mildred declared; "it's where you got your money." Mildred didn't mind: "I'd hate to go through it now, but then I didn't know anything different." Housework was a lower priority than farm work. "I never was crazy about housework; I done it when I could, but I worked outdoors mostly." She found household chores "awful hard" and monotonous, "the same routine day after day." "Maybe it didn't get done much. And we never had nothing."

When I asked Mildred how she and her husband made decisions, she replied, "Well, I guess when we needed" things for the farm, "we went and got them." She did not remember talking things over with Thomas. She left the business decisions up to him "pretty much. He knew what he wanted, and he just got it." Mildred's initial use of the first person plural "we" is significant; although he made the purchases, she shared the conviction that whatever the farm needed came first. Her husband left decisions about the house up to her. "I just had to ask him for what I wanted. I knew just about what I could have and couldn't have, and you had to spread it out. He used to give my money for groceries." Although Thomas provided the cash, it was hers to spend as she saw fit; he trusted her to be thrifty.

Although Mildred routinely participated in farm work, her husband never helped her in the house. "A lot of men would help do things for you, but he wasn't that kind. My father didn't do nothing, either; that wasn't a man's job"—or so these men thought. It was harder on Mildred than on her mother, however, because she had no daughters. The Martins' youngest son eventually took over Meadow Creek Dairy. He worked with them for wages at first, but in 1957 he bought the farm and paid his parents wages for their work. The businesslike relationship between the Martins and their son ensured a successful intergenerational transition of the commercial enterprise.

Mildred Gale Martin's labor was integral to the dairy farm operation throughout her long life. The matter-of-fact way in which she acknowledged her husband's greater say in decision making is balanced by her self-confidence and the authority she exerted over their son. Still, her story exemplifies the fact that active participation in outdoor labor and involvement in marketing did not automatically translate into decision-making leverage, just as Clara McGregor's story demonstrates that a woman who worked mainly in the house was not necessarily marginalized.

"A Partnership Is Hard"

Louisa Woods Brown, who was born in 1907, married a man who ran a farm in partnership with his brother, as his father and uncle had done. Her parents had bought a farm when she was a child, but they were always poor. Her husband and brother-in-law's farm was a large-scale, specialized, and successful commercial operation. The implicit and explicit contrasts between the two families resonated throughout her narrative.

When I interviewed Louisa, her husband, Peter, had been dead for less than a year. Her youngest child, who was disabled, still lived with her. Two daughters had married farmers and were doing well, one in Whitney Point and the other in the Midwest. Her two sons were running the Brown farm in partnership. The brothers worked well together. Louisa explained that they "complement each other, because one is a good mechanic and the other has a real business head on him." They have continuously expanded the operation; at the time, it was the largest retail and wholesale business of its kind in Broome County.

Louisa was very knowledgeable about the history of the family farm; indeed, she knew more about her husband's ancestry than her own. His mother was descended from the Scottish immigrants who lived on Mount Ettrick, and his parents farmed the land she had inherited.[7] They had seven children, but their eldest son had died of measles and Peter's twin brother had died in the 1918 influenza epidemic while they were attending a winter course in agriculture at Cornell. That left two sons, Peter and Douglas, and three daughters, all of whom married farmers who lived in the neighborhood.

In 1930, when Louisa married Peter, she was twenty-three and he was thirty-three. They decided to build a house of their own on the Brown farm "because we wanted a family." They put in a wood-burning furnace and an electric stove, and they added on to the house as their family grew. Right from the start they provided housing for the partners' hired hand. Douglas, who had not yet married, lived next door with the brothers' widowed mother and her unmarried sister Agatha, whom Louisa called "Aunt Aggie." Aggie received payments from her siblings for her part of the family's land. She also "used to work out a lot" doing housekeeping and childcare for families in the open-country neighborhood. Louisa appreciated her assistance: "She used to take care of my little ones. She loved them so, the babies." Louisa needed help with childcare because she spent long hours working on the farm and in the business.

Louisa "found a way" to combine childcare with farm work "because I was always out there doing something." Indeed, she never discussed her work without explaining how she ensured that her five children were cared for. When she first came to the farm, they sold a larger quantity and wider range of produce than their own farm could supply. "My husband would start out at three o'clock in the morning" on buying trips. "Lots of times I'd go with him, and that's where Aunt Aggie would come in and stay with the children. I did a lot of the driving for him, in the truck....We'd get home maybe at midnight." When Louisa was home she'd work in the berry fields, sort produce, or staff the farm stand. She took the youngest children outside with her and put them in the car when she ran errands. "I was with 'em most of the time, you might say, more than their father was. Because he was gone so much. They'd be lookin' for him at night, and he'd have one on each arm when he was eatin' supper, because they wouldn't be asleep....Sometimes, well, lots of times he was too late to get home; sometimes he wouldn't see 'em from one week to the next, hardly."

When the children were young, "I did have a hired girl in the summertime, when I was working in the berries. Aunt Aggie wasn't always here." Louisa would hire a girl from a neighboring farm, who "did the cooking and watched the kids and did the washing." "I never had anybody in the wintertime." The children all "worked, when they got to the age where they could do anything. They each had their projects; they had chickens or pigs or a strawberry patch; they had things to earn their own money." She spent so much time outside that the housework "probably suffered."

To add to her income, Louisa took in foster children for $50 a month. That "wasn't much, but it helped out; times were rather hard." "I probably had as many as ten or twelve over the years." The money paid for "little things that I would like, extras, you know, that you couldn't get out of what income we had, which wasn't very big. To make a go of the farming business, you had to be pretty careful." Sometimes she spent her earnings on things for the house: "a new bedspread, or some curtains....I always liked to keep things nice, as much as I could. And maybe towels or bedding or something of that sort, *necessary* things." "The welfare" gave her an allowance for the foster children's shoes and clothing, but by sewing their clothes along with those for her own children she could use the money to buy shoes for all of them.

In addition to the farm's main commercial operation, they had a small dairy, keeping a dozen Jersey cows and selling high-quality, sweet cream butter. When Louisa came to the farm, Douglas was running the dairy with some help from Peter. Louisa soon took her husband's place, and gradually she assumed primary responsibility for the operation. They always had milking machines, which lightened the labor. Louisa "used to...milk about five cows morning and night"; Douglas did the rest. After the barn burned, Peter wanted to rebuild and expand the herd because he hoped that one of their sons, who was interested in dairying, would take over the operation eventually. But even with twenty Holsteins the dairy was too small to be profitable, so they sold the cows.

The adjacent farm was owned by other members of her husband's mother's family. Louisa's mother-in-law's brother, and then his two unmarried daughters, had inherited it. By the time Louisa married into the family, that farm was operated separately, in conjunction with another farm that a third brother had nearby. The fourth brother had been part of the operation at first, but he and his wife got into a dispute with his brothers and sisters, so they left and started their own farm elsewhere. This series of events served Louisa as a cautionary tale about what could happen when siblings' spouses came into conflict: relationships were ruptured, and farm partnerships dissolved.

Peter and Douglas held the land in common and never considered dividing up the operation. "They never...never, I don't think, had that idea at all. Maybe I did, but he [her husband] didn't. It was hard; a partnership is hard sometimes. I think maybe it's harder on the wife that comes into a family like this, because...well, it just is. I mean, you're working for two men, instead of one, really! And he didn't have a wife, and I was doing a lot of the work that a wife...I used to do his washing, and I quite often cooked for him, and did things like that."

Louisa was knowledgeable about the business because "I did a lot of the book work." She participated in making some decisions, but "I think they made more decisions between themselves....I remember once, they borrowed some money and didn't tell me, and I was pretty irked about that. I don't think they did it a second time. Because I felt that I should know when they borrowed money, you know. But I can't remember having any big problems about it." Louisa recognized that she was not a full partner but, rather, a partner's wife. Yet "I had a sense of duty; I felt that I had to work hard" to make the farm successful.

When we discussed the process of intergenerational transition in detail, it became clear that one of Louisa and Peter's daughters might have wanted to come into the farm operation with her husband. She had married a man who worked for the Browns, and they lived in the tenant house on the farm. But at that time, "There wasn't enough assets for him to come into the business and have our two boys in too." The youth "always wanted to go in business for himself." So his father, who was a skilled tradesman, invested in some land, and the young couple moved to their own farm. Later, the Brown brothers contributed capital to their sister and brother-in-law's operation. They also contributed capital to the other sister's enterprise when she and her husband decided to move from the land he had inherited from his mother to better land. In effect, the brothers bought out their sisters' shares in the Brown farm.

The contrast between Louisa's own family and the Browns was central to the way she told her family's history. Her father had been a railroad worker, but "the doctor told him he would have to get out of the city and away from the railroad smoke or he would have consumption." When she was about two, "they bought this farm over here in West Chenango" with the help of a great-uncle. Louisa recalled hearing that they had a horse and one cow when they came. "It wasn't easy" to make a living with little livestock and no equipment. Her mother, who had grown up on a farm, "always helped do the milking and pick the berries," as well as cultivating a large garden. "We had enough to eat," but she and her two younger siblings "went barefooted to school." Louisa started high school and boarded in town, but after two years she had to come home to help her parents. "I can't remember ever having very much, when I was home. But I think we were happy." Her father started out with 75 acres, but "he needed more land to make a go of it," so he bought the adjacent 56-acre farm.

Louisa recalled planting potatoes and corn, as well as picking berries. "I remember one day the horse stepped on my foot, and I thought, well, I haven't got to help plant those potatoes today." But her father said that "pushing the potato eye down into the soil 'will cure your foot,' so I didn't get out of planting potatoes!" She laughed ruefully as she remembered her father's reaction. Other memories were more positive: "I used to drive calves, used to train 'em and drive 'em when I was a young girl." Her grandfather, who lived with the family, fixed up an old harness so she could hitch them to a cart. She recalled with amusement how one calf "came up behind me and bunted me right over" the fence "into the blackberry patch." She loved horses, especially her grandfather's Morgan. Her father had two huge draft horses. "I used to drive the team on the rig and rake up hay, and drive the horse to pull the hay up in the mow." Her siblings did the same range of farm tasks; nobody said anything about what was or was not proper for a girl or boy to do. In the neighborhood, "Not all the women worked on the farms like we did. But a lot of them did."

Louisa also "cleaned house and mopped floors and brought in wood and carried water. We didn't have any indoor plumbing"; "we had a well just outside the door" with a pump. "We had to carry all the water for washing and dishes and everything. I remember getting in a fight with my mother one day; she said to get hot water and do the dishes, and I said cold water'll do that just as good as hot water. I used to have to do the dishes before I went to school in the morning." Louisa helped to cut up stove wood with the buzz saw, which was powered by a gasoline engine. The washing machine, however, was hand-powered. Her parents installed carbide lights in 1930, before the neighborhood got electricity.

The inheritance of the Woods family's land created problems between Louisa and her sister. Their brother was never interested in farming and got a job elsewhere. When her father stopped farming in the early 1940s, he "sold, or pretty near give," the larger farm to her sister and moved to the smaller farm. Evidently their mother was responsible for this decision: "When they sold the farm to my sister, my father didn't have much say in it; he wanted to get more money out of it. It left them in a bad way; they had very little to get along on." After her mother died, her father left the smaller farm to Louisa. "There was bitter feelings there," she recounted. "I told my Dad, when he told me he wanted me to have the farm, I said, you'd better not, it's going to cause a lot of trouble. And it did. My sister just couldn't accept it. She had gotten the other farm for just a fraction of what it was worth, and still she was very bitter because he gave me the other farm. I felt badly about it, and I'd rather he wouldn't have done it, but...." Giving her the smaller farm did make the inheritance fairer. "But I still didn't want that feeling between us." "My sister had told me, she said, 'I always wanted that place.' But she sold the other place and got a good price out of it, so she shouldn't have felt that way, I don't think...." Louisa's sister and her husband did not farm. "You see, my Dad wanted his land to be farmed, and he felt my boys would go on with it, which they are....They've got strawberries over there now, and...they've got a lot more of the land plowed up. It's good land, and they'll make use of it." To Louisa's sister, however, the land was an asset to sell or a place to live. After she sold the larger farm, she moved a trailer next to their father's house. Louisa "had to give her back enough property to put her trailer on. So, of course, that was hard. Not for me, but for her. But it was hard for me to feel she was that bitter. And there was nothing I could do about it, not a thing."

Louisa Woods Brown's narrative emphasizes the contradictory pulls that a woman who married into a farm partnership experienced. On the one hand, she had to commit herself to the success of her husband's enterprise and to ensuring intergenerational succession. On the other hand, she had to be careful not to offer opinions where they were not wanted or to disturb the working relationship between her husband and his brother. Louisa accompanied Peter on purchasing and marketing trips before their children were born. Then she made a place for herself in the dairy and berry patch. The older women in the Brown family helped by doing childcare and housework when she was busy outdoors. By working independently in farm operations of her own, as well as by taking in foster children, Louisa contributed to the family budget, relieving the farm of financial burdens and raising their standard of living. Her awareness of the difficulties that had plagued her husband's relatives on the adjacent farm and her regrets about the conflicts over inheritance within the Woods family made her especially sensitive to interpersonal relations. Her realization that working together harmoniously and avoiding conflict were crucial to intergenerational succession enabled her to facilitate her sons' accession to the business and her daughters' establishment of their own farms.

"'You Keep Out of It' Was More The Idea"

Mary Connolly and Esther Bond married into the Stewart family. Their husbands, Hugh and Gerald, farmed in partnership with their father. James and Ella Gordon Stewart had established a specialized market garden that became large enough in scale to allow both of their sons to stay in the business. Indeed, the farm operation was labor-intensive, requiring round-the-clock supervision of the greenhouses and extended marketing trips, so having several family members working full-time was essential to the profitability of the enterprise. According to the daughters-in-law, however, it was not a foregone conclusion that both sons would go into the business. The fact that they came of age during the Depression made a significant difference. So did their decisions to marry and start families.

Mary Connolly and Hugh Stewart, the oldest son, were married in 1934, after he graduated from Cornell. She was twenty-four and he was twenty-three; they had courted for two years. Mary had completed normal school and taught in an impoverished rural community near the Pennsylvania line. Although she had grown up in a working-class family in Endicott, she had no hesitations about marrying a farmer. "I thought farmers were very nice. I had lived in the country when I taught out there, and I don't know that I thought that there was any stigma attached to that....I had a lot of respect for farm people." Mary "wasn't afraid of being overworked." Indeed, she said, "I thought it might be nice to get out and do some of the things" that farm women did outdoors. "But it was different; it was more specialized by

the time I was here, anyway. It wasn't even like living on the farm, the way"
the Stewarts lived. "That's what the girls in town said" when they met her
husband-to-be: "'That's what you call a gentleman farmer'. Not the way we
think of a farmer being uncouth, or not well-educated or anything. I remember my brother saying, 'Wow! Those boys certainly have good manners!' Of
course, they stood up when my mother came into the room and all that; my
brother didn't do that." Mary "didn't think of farming as a hard life because
they had so much more, the Stewarts did, than we did at home."

In retrospect, Mary wondered whether farming in partnership with their
father was the best thing that Hugh and his brother could have done. It
supported them all, to be sure, but it "didn't necessarily make them happy."
"Their parents sort of expected them to stay in the family business....Hugh
was the only one, really, who had any interest in" market gardening; he
studied horticulture at Cornell. "As I remember it, Gerald was very interested in dairying. But there comes a time when you've got to have money
to live on, and it's a good thing if you can inherit your father's business."
"I think that was his parents' dream, that someday the boys would work with
their father. Of course, they were wise; they didn't say you have to do this or
anything. But they never encouraged Hugh to take any teaching courses to
teach Ag., so he didn't. He always planned to come back on the farm."

Mary continued teaching in Endicott until 1938. She had not planned to
have a career after she married; "most girls didn't expect to, at that time."
But it was five years before she had her first child. She made good money
and appreciated the financial security during the Depression. For the first
two years of their marriage she and Hugh lived in an apartment in his parents' house. Then they built a house for themselves. "All my pay, as fast as
I could take it down to the bank, there it went, saved up for this house."
After that they didn't need her income so much, and Mary planned to stop
working. But "I guess things weren't going too well, bills were high and so
on, so I said, why don't I teach another year?" Once the children started
coming, however, she had more than enough to do at home; eventually she
bore eight children.

Mary never became involved in the Stewart family business. "Oh, I might
go out to the greenhouse with him or something, but I never did much work
on the farm at all. Not because I thought it was beneath me, or anything like
that. But they all had their own jobs to do. And they had women working
in the packing shed; they didn't need me there. I did go around with Hugh
some of the time just for the fun of it." Her husband did not want her to
work in the family business. "That was never any of the Stewart men's idea.
I think, 'you keep out of it' was more the idea."

In Mary's eyes, the clear gender division of labor in the family was a
mark of their class position. Mary had grown up in an urban working-class
family. Her father was an alcoholic, and when he was drunk her parents
quarreled. Her mother always had to work full-time just to keep a roof

over their heads, and the children got jobs as soon as they could to feed and clothe themselves. The Stewarts were very different. "I think in their generation, that particular type of man had a lot of respect for women," Mary explained. "And their idea of the way a woman should be—I don't mean that a woman's place was in the home, exactly, though it seemed to be that way. But they seemed to respect a lot." The men made sure that their mother and their wives did not work as hard as some farm women had to do, at least not outdoors. "She might work from dawn to dusk in the house, or doing women's work, you know." But the Stewart sons did not need their mother to earn money. The family had many household conveniences and labor-saving devices, and during the Depression household help could be hired very cheaply.

Esther Bond married Gerald, the Stewarts' younger son, in 1936. She was not quite eighteen, and he was twenty-two. Gerald had been working with his father and saving money for college, but he decided to get married instead. "We lived very lean until we got the business built up," Hugh explained. "I was making $10 a week when I got married, and after that I got $15 a week and a place to live." To Esther, living with the Stewarts was not a hardship. She was born on a small farm in Pennsylvania, but her father gave up farming to work for his brothers in construction, and they moved around a lot. "Then the Depression came and wiped them out financially....My father salvaged what he could," and they came to the Nanticoke Valley. "They were looking for a place where they could have a little farm operation and be self-supporting until they found something better." He never found another job, so her parents spent the rest of their lives on the edge of Maine village. They kept "a cow, a pig, some chickens, and some sheep" on their small plot of land. "Mother worked extremely hard. She made the garden, and she canned. She helped with the chickens. Mother churned. She made our clothes. I can even remember her making soap. Mother used to cook on an old wood stove. She used these copper wash-boilers, and she did her canning in that." "Those were very lean years. We were very much on a subsistence level." Esther "never had any qualms" about marrying a farmer. "I thought that when I married, I married out of poverty."

Esther, like Mary, never worked on the farm, and she did not become involved in the family business until after her children were grown. Although she had helped out at home, "this was a different situation; there were a father and two brothers involved." Gerald continued, "The worst thing that can happen with brothers in combination like this, trouble, is with the daughter-in-laws, or the in-laws. That didn't enter in; we had practically no problems." He explained that their caution was based on watching what happened in other families. "Say if there's two brothers; the two brothers don't have trouble, only it's caused between their wives. I've seen that so many times." Brothers were used to working in tandem, while their wives came from other families and might see things differently. The women would "get in each other's hair" or think that their husbands "weren't getting a fair deal." Gerald declared emphatically that "there really was no place for" his wife "on the farm"; "the business at the time was too big."

Still, Esther had a voice in family decision making. When I interviewed her separately from her husband, she said that both Gerald's mother and her own mother were "submissive" toward their husbands. "I think that years ago girls were brought up to be submissive to their husbands, and that was their role in life." Esther "was brought up that the husband was the head of the house," but she rejected that idea. Her own marriage became a more equal partnership. "I think that I started taking more of the initiative in our life, when I wanted to have things.... Well, Gerald would come home sometimes from work, and he would find the wallpaper torn off the wall. I knew that if I tore that wallpaper off, I would have to have some new to put up. So, if I wanted something done, I would go ahead and do it."

Indeed, Esther's preference ensured that they stayed on his parents' farm rather than starting a farm by themselves. At first Gerald said, "I don't think I ever thought of doing anything else." But, Esther reminded him, "There was a time in our life when we were a little undecided about...whether we were going to stay in the business or we were going to branch out for ourselves. Your Dad, I think your Dad at that time was trying to encourage us" to start their own dairy farm. Gerald acknowledged that he had forgotten about that. "We looked at the situation," Esther continued, "and I think I sort of took the initiative there and tried to encourage you to stay here with the business. By doing so, we could take our savings and build a new home, instead of trying to strike out by ourselves." Gerald added that "my Dad thought it was very impractical to build a beautiful new house. That was why he thought we should go buy a farm, put the money into a farm instead of just having a nice house. That was the basic issue." Gerald went along with his wife's preference. It made no sense to his father for a farm family to put their savings into a house rather than productive property. It seems clear, however, that Esther was worried about whether they would make a go of it by themselves; after all, her parents had left their farm and then lost the family business in her childhood.

Working with their father did not allow Gerald and Hugh the autonomy that most adult men enjoyed. The older man "never divulged any information about the finances of the business. He kept those books religiously to himself." Laughingly, Esther declared that he "would sooner have walked bare naked into a room full of strangers" than open up the books. He was even reluctant to show them to the IRS. The sons were not formal partners in the business as long as their father was alive. "For years, we were classed as workers" for Social Security, Gerald explained. "We worked as partners and we *were* partners, but legally there was no such document ever made." Instead, there was an unwritten "gentlemen's agreement" that they would inherit the farm. When Esther and I discussed the matter privately, she explained that the older man excluded his sons from effective participation in farm decision making, just as he excluded his daughters-in-law from involvement in the enterprise. He kept his sons "living on a shoestring." Every year, after he reckoned up the profits, he would give them a bonus, but "we were supposed to put that right in the bank." It was difficult to manage. The men's wages were low, although they got some food supplies from the farm. Their father occasionally purchased things their families needed instead of giving them cash.

Esther held a job off the farm for a year before her last child was born. It made financial sense for her to go out to work. Gerald explained that "she made considerably more money than we were paying" the women they employed. "At that time," Esther continued, "the idea was to make as much money as I could." When their children were in college, she went to work in the Singer-Link and General Electric factories in Binghamton. "Later, if there was a shortage of help" on the farm, "I'd go over and help out." Eventually she took over the bookkeeping from her brother-in-law. She continued working in the office when their son came into the partnership and more than doubled the size of the business. "After all, who can you trust like your mother when it comes to handling money?"

The women who married into the Stewart family accepted the clear gender division of labor that was created by their husbands' partnership. They also recognized that getting along with their in-laws required them to stay out of the Stewarts' business affairs. It was easy for a family to blame any differences that arose among brothers on the jealousy of their wives, and both husbands and wives resolved to avoid furnishing the slightest pretext for conflict. At the same time, neither woman was willing to be "subservient" to her husband. Crucially, both worked off the farm when their families needed the income. In distinct ways, each asserted her power within her marriage and promoted shared decision making in the name of "togetherness." The reciprocity that in the older generation took the shape of "respect" was enacted in the next generation as an easy give-and-take between husband and wife. Mary and Esther did not allow the class privilege that exempted them from farm labor to undermine their position. Both worked off the farm and thought of themselves as their husband's partners, not as marginal to a partnership of men.

"I Wanted to, and We Almost Had to Work Together to Get Things Done"

Hattie Bieber Smith and her husband, Edmund, had a partnership that was unusual when a man farmed in partnership with his family. The Smiths negotiated a relationship with his father that allowed them a significant degree of autonomy. In their own farm operations, they acted like equal partners, and Hattie often took the lead. Farming also enabled them to follow their own ideas rather than conform to what they regarded as stifling, narrow-minded, small-town culture.

Hattie established a close working relationship with her husband while avoiding conflict with his parents. When they married in 1926, she was twenty-four and he was twenty-seven. They set up housekeeping in a renovated shack on the Smiths' land. "It had been used to store everything from hay to mice to…no windows…oh, it was a mess. But we came in here by ourselves instead of going down to the farmhouse, because his parents still lived there." Hattie and Edmund especially valued being able to spend so much of their time together. While Edmund went down the hill to the barn twice a day to milk, Hattie cultivated a large garden, built up their poultry flock, and raised feed grains and hay. In effect, the family operated a small-scale, market-oriented farm of their own alongside the commercial dairy farm that Edmund operated with his father. Through their labor on the land, they explained, they were able to live in their own way.

Significantly, Hattie and Edmund always talked with me together. In two long interviews, husband and wife took turns as narrator, one interpreting a story the other had begun or supplying more details about the subject under discussion. Both used *we* much more often than *I, he,* or *she.* Each spoke of the other's experience with the assurance that comes from a shared understanding. Although they discussed their occasional differences of opinion as well as what they agreed about, they did so openly in one another's presence. They concurred both about farming and about childrearing. Critical of the authoritarianism they observed in other families, they taught by example rather than precept and expected good behavior rather than punishing bad. Being home was most important to Edmund. He loved having midday dinner with his wife and children. He enjoyed eating fresh food, exploring the natural world, and talking over national and international events. In the evenings, Hattie read out loud to everyone from books they borrowed from the library, or they listened to the radio.

The Smiths did not socialize with people who did not share their liberal views. Although raised as Christians, both were what would then have been called "freethinkers." The family rarely went to church, but spent Sundays rambling, reading, and listening to classical music. "We aren't firm believers in any orthodoxy," Hattie and Edmund remarked; "we have our own ideas, and we talk a lot." Those ideas included a firm belief in economic cooperation among farm producers and with urban consumers. The Smiths were "progressive" farmers in both senses of the term: they experimented with innovative agricultural techniques, and they were active participants in radical farmers' organizations. Most remarkably, they were strong internationalists, in later years traveling to Russia and China to meet country people on collective farms.

The Smiths were also deeply committed to education. Hattie had taught for five years before she married, and the district school did not measure up to her standards. So they enrolled the children in a better school in the nearby city and drove them down every day on the newly paved Farm to Market Road. All their sons and daughters went to college, although paying tuition was always a struggle. They passed their own farm and a share of the Smith partnership on to one son. Another became an agricultural scientist; the other sons and daughters all entered urban professions. The Smiths felt free to make choices about how they lived because the farm provided them with a modicum of independence from the rest of society. Hattie took for granted that she was an equal and respected partner in everything. To her, the expression of shared values was the essence of both personal agency and family solidarity.

As our conversation flowed back and forth between Hattie's life history and the history of the Smith farm, it became clear that the decision to develop their own farm operation, which relied centrally on Hattie's labor both indoors and out, was the result of her personal preferences as well as economic necessity. Edmund had always assumed that he would rejoin his father on the farm after going away to high school. "It was a two-man farm and they needed the extra hand so I stepped in." His father's brother had died, and he was the only son. Edmund worked on the presumption that he would inherit the farm after his death, but they had no formal partnership agreement, "not even a verbal one." Hattie explained, "He and his Dad got along just like that"—she crossed her fingers to show how much they depended on one another—"and that was all there was to it."

The younger couple did not get a regular income from the joint operation, not only because of the vagaries of weather and prices but also because of their lack of a definite financial arrangement with his father. "If they came home with money from selling their products, that went for taxes first. I have known when we had three or four dollars a week to buy groceries with." Edmund worked off the farm on the cooperatively owned telephone line each fall to earn cash to pay his life insurance premiums. "Otherwise, he had some of the money that they came home with from their chickens and butter and things.... No, there was no wages or anything. We talked about it. We'd sit down and say, come on, Daddy, do you want this to go on, and so on. Well, yes, he wanted to stay; the farm had been in the family for so long. And Granddad was getting so much older, and he wasn't wanting to talk about things of that kind."

To make a living and provide for their children's education, then, it was necessary to develop separate income-producing operations. They always had a poultry flock. "To begin with, up here, ourselves, it was just for our own family. But later we raised broilers and dressed them ourselves." "Granddad made butter and had a route, with customers in town," so he sold their eggs and poultry along with his farm-made butter. They kept the hens out behind their house. The children took turns helping Hattie feed the chickens. Hattie washed the eggs "if they absolutely needed it," weighed and graded them, and packed them into crates. They sold broilers "all over Binghamton and Johnson City....Ready for the table, they were. I raised some parsley and put some parsley under their wings. Some of the women would say they'd buy the chicken to get the parsley," she said with amusement. "The whole family helped do the picking, but I did all the drawing myself," right in the kitchen.

Hattie also worked in the fields. She began by "driving the horses to take the hay off with a big fork in the barn." Soon "I went out and helped rake hay," even when their youngest child was small. Short of capital and without a tractor of their own, they improvised. Hattie laughingly described her technique: "We put the rake, not a drop rake but a side-delivery rake, behind the family car. Put the children in the back seat looking out the back window at the rake while I would rake up the hay with the car." As she drove across the bumpy meadow, the children would warn her if she missed a windrow or the rake flipped over. "Over the years," Hattie said later, "I baled all the hay. And in the spring I enjoyed working the ground; I used to do that day after day. Not plowing, but discing, working the ground with the big disc on the tractor. No, I didn't plow, and I didn't mow grass; but other things, yes. And I loved it."

She preferred working outdoors to doing housework, and she knew that it was more important to the family's living. "If today was...time to get in hay, you did that and you didn't clean the living room. You did what had to be done. You got the meals....We've talked about it since, how many meals there would be in the fifty-six years. You did what had to be done to keep the family going, but then you would do the jobs that came along. When it was time to thin the carrots, you thinned the carrots, if you didn't dust, you know, things of that kind." Hattie made no effort to keep to a conventional weekly schedule for housework; by necessity and by preference, farm work took priority.

When the Smiths moved into their house, "we drew water by a rope and a pail. Daddy carried the old water out; we had a ditch" outside the kitchen door. Soon they improved the house. First they installed a hand-operated pump. They purchased a battery plant to run the lights, the pump, and their washing machine in 1930, when their twins were born, rather than waiting until the power line arrived. At that time, Hattie had four children in diapers and was doing laundry every day. "We had a man who would come through with a truck of bread...and he asked me one time, 'Mrs. Smith,' he said, 'do you take in washings?' I said, 'Take in washings? I've got enough to do!' 'Well,' he said, 'your line is always full!'" Eventually they modernized the kitchen. But Hattie had the impression that they had fewer household conveniences than other farmers in the vicinity.

Hattie raised vegetables and canned them for winter; one year, she recalled, she put up 500 quarts. But "I bought bread more often than I actually baked it." "For one, it was cheaper"; at that time, wholesome bread sold for

10 cents a loaf. "And there were six lunch boxes to fill when the children were all in school." Hattie didn't calculate whether she actually saved money by buying bread rather than making it herself. "No, we very seldom figured that way"; they focused instead on just getting by. She was not accustomed to counting the cost of her own labor, either. Rather, she weighed the time it would take to do household tasks against the farm work she could accomplish during that same time.

When Edmund was out of the room for a minute, Hattie emphasized to me that he had always left it up to her to decide what work she would do on the farm. "I can't remember that he actually would" ask her to do anything on a certain day. They might talk over what needed to be done. "If I said no, that would have been it; that would have been it. But I wanted to, and we almost had to work together to get things done." Hattie knew what the farm needed as well as her husband did. "And he has sat and helped me snip a bushel of beans or something like that, too. Not leave something that really needed to be done, but he would hustle around so he could help me." He never expected to come in from the barn and rest while she worked. "And if I didn't have the meal right ready or something, he'd say, 'What's the hurry?'...No, he's patient with a capital P."

Edmund's account of the farm turned entirely on the work itself; he never so much as mentioned any tasks being assigned by gender. When I asked him directly about the allocation of tasks in his parents' day, he said, "I think it was more a matter of timing. There was a time when, comparatively speaking, they had a lot of cows, and everyone that could help, helped milk. And later on they had a lot of chickens, and at that time whoever was available helped with the chickens." Hattie interjected that his stepmother and his sisters never went to the barn or did much work outdoors. Edmund agreed that was true, but declined to speculate about the reasons. Hattie continued, "When I did it, it was because, partly it was there to do, and I wasn't told to do it. I wanted to do it; I had fun. It was a job, I don't say that it wasn't. And sometimes there was inside work and all that to do. But...it wasn't a 'have to' sort of deal."

Edmund described his father as among the progressive farmers in the vicinity. Deeply interested in scientific innovation, he experimented with putting lime on the acid soil and with fertilizing meadows as well as tillage land. The Smiths were among the first to set up a silo to ferment their own corn for cattle feed and to use pen stabling. His father was always active in the Grange, serving as a county and regional officer, and he joined the Farm Bureau soon after it was formed. By the time Edmund was working with him, they were milking 12–15 cows. Unusually, they did not sell fluid milk but produced high-quality, sweet cream butter that they sold directly to consumers. They had a mechanical separator and cooled the cream in the well. Then Edmund, rather than his stepmother or, later, his wife, churned, washed, and packed the butter. It was Edmund's job to deliver it to their customers in town. Because his stepmother did not want to be involved in the dairy operation, Edmund brought the milking machines up to their house, where Hattie washed and sterilized them. Neither she nor their children went down to the dairy barn regularly to milk either. That was as much an interpersonal as an economic matter, however; William's wife did not like children, so they kept their distance from her.

After his father died in 1941, Edmund and Hattie took over the whole farm. Edmund paid off his sisters, although their shares were not equal to his own since by then he had devoted fifteen years of labor to the enterprise. The Smiths concentrated on dairying and poultry, rather than continuing the diversified operations his father had conducted; they got rid of the sheep and stopped selling apples and potatoes. Most important, they started sending fluid milk to a dealer with a processing plant in Oakdale and retail customers in Johnson City. The dealer bought their eggs as well. That relieved them of the time-consuming necessity of "running from here to there to way over there" to reach their customers. Selling fluid milk required them to improve the barn, but they did not expand the operation until their youngest son joined them in 1955.

The Smiths did not necessarily expect any of their sons or daughters to farm. "I don't know that I ever hoped anything about it," Edmund explained; "I figured they had to make their own decisions." The older sons and daughters established independent careers. The son who became their partner valued the quality of life and closeness to nature more than financial success. "That reminds me," said Hattie; "we had a beautiful basket of apples sitting here once, and an insurance man came, and he remarked on the gorgeous basket of apples: 'I suppose you've set those aside to sell.' I can remember what Daddy's comment was. 'Oh, no,' he says, 'those are for my family; I always pick the best out for the family.' The man was aghast. Because we had nothing...I say we had nothing, but we always had plenty to eat and we always were warm, but we *had* nothing." Edmund added the moral of the story: "That's part of the pay you get, is using the best of what you grow."

Hattie Bieber and Edmund Smith worked as partners in their own farm operations while he continued working with his father in the family enterprise. Like Louisa Woods Brown, Hattie maintained a degree of distance between her own work and that of the family into which she had married. Yet, to a much greater degree than Louisa, Hattie shared labor with her husband and children. She was responsible for the land they cultivated and for their poultry operation; she enlisted her children from any early age, and her husband was always remarkably cooperative. The values of independent thinking and neighborly cooperation that Hattie and Edmund shared linked them with other country people. The mutuality that characterized their own working relationship served as the basis of their political-economic as well as their familial ideals.

7

Wage-Earning and Farming Families

In the Nanticoke Valley, the consolidation of larger farms was accompanied by the proliferation of smaller operations, and class differences between more and less successful farmers widened. Between World War I and World War II, as the scale and degree of specialization of commercial farm operations increased, women's work became less central to these enterprises. Yet women's active participation in income-producing labor, whether on or off the farm, was a fundamental feature of working-class families who combined farming with wage-earning. By the end of World War II, this pattern prevailed throughout the hills, and full-time farming families clustered in the valleys.

Most of the immigrants who bought farms in the Nanticoke Valley kept some family members in the urban labor force while the others worked on the land. At the same time, many of the poorer native-born families also sent people to work in the city. Like their counterparts from the European peasantry, these families struggled to maintain a foothold in the country by combining wage-earning with farming. In some families, men held off-farm jobs while women cultivated the land; in other families, these roles were reversed. Nothing except personal preference—not even prior experience in agriculture—controlled which pattern families chose. Women negotiated with their husbands about the family's labor arrangements.

"Yes, She Worked out in the Fields"

The Dudzyks, who bought land in the town of Maine in the 1920s, were among the many immigrant families who moved from the manufacturing

cities of Broome County to the Nanticoke Valley in the early twentieth century. Wilma Dudzyk Kowalski was born in 1919 in Milotice, a town now
in the Czech Republic that was then part of the Austro-Hungarian Empire.
She identified herself as "Moravian" and her people as "Slavish."[1] Her father wanted to farm full-time, but he was never able to do so. He continued working in the tannery owned by the Endicott Johnson Shoe Company
(which was commonly called E-J) while her mother and the children did
most of the farm work.

"I came from the old country," Wilma began. Her father left for the
United States and worked in Endicott for two and a half years before he
could bring his wife and three surviving children. Wilma was just four when
they arrived in 1924. Three more children were born here. The only relative
they had nearby was her father's older brother, who had come before the
war and had also worked in the tannery before buying a farm. After the family had lived in Endicott for six months or a year, Wilma's parents bought
a farm near Bradley Creek Road adjacent to her uncle's. Several other immigrant families lived nearby.

"My mother had cows, and she used to milk the cows, and she made
butter and cottage cheese, and she'd take it downtown and sell it from door
to door in the Slavish part of Endicott." She used a mechanical separator: "I
remember washing it, because we had to wash it every night." Wilma herself
never learned to milk. "I helped out on the farm, but we didn't have that
many cows. Maybe we had five or six cows, but we didn't have a herd of
cows where we milked them and gave the milk to the creamery." They also
kept chickens and sold eggs.

Wilma recalled that "we'd hay and bale oats and dig potatoes and
weed" the vegetable garden. Her mother worked outdoors as well as in
the barn, and "us kids would help when we were out of school. Yes, she
worked out in the fields.... My father plowed, but she'd plant the corn and
the potatoes." "We all hayed. In the old days, they'd cut it and let it dry,
and then rake it up. We'd have to go along and put it in piles. My father
would come along with the wagon and we'd throw it up on the wagon
for him to take it into the barn." Wilma's mother worked in the hayfields
with her father, riding the rake and turning the hay. The men cooperated to
harvest the grain. "My uncle and my Dad, when they had their oats ready,
they would get a bunch of men and they'd do one field and then the other.
My mother and my aunt would get together and they'd have a big meal
for them."

Except during haying and harvest, her father was often away from home.
"When we first came here, he used to stay downtown all winter long. I think
he came home on weekends, maybe. But later on he got a car and he'd drive
back and forth." He still rented a room in Endicott, but during the summer
he would come home every day because he was needed on the farm.

Figure 8. Ina Tymeson with milk cow and calf, Maine, New York. Undated photograph by Lee Loomis. *(Courtesy of the Nanticoke Valley Historical Society.)*

Wilma thought that her father "used his pay for the farm, for the mortgage," and "to buy things for the house," including a treadle sewing machine so her mother no longer had to sew the children's clothes by hand. The house had no utilities or plumbing. "We didn't have no electricity, we didn't have no inside toilets. We had a sink and pump in the kitchen, but that was all." Wilma's mother "never was used to having any" household conveniences, "so she got along with what she had." They had electricity installed after 1940, when Wilma and her baby were living with them; her father bought a washing machine with a motor.

Her mother used the proceeds from the dairy products, eggs, and vegetables she sold in Endicott "to buy groceries." Wilma had the impression that, when it was a question of making decisions about the farm, her parents "talked it over. I think my mother had a lot to say with it." For example, she remembered that when they sold a cow to an Italian farmer her mother was reluctant to take his check. Wilma's parents paid off the mortgage and later bought more land above theirs on the hill.

The Duydzks never managed to farm on a large enough scale to enable them to devote their energies entirely to agriculture. "There was a lot of them like that, where they had big dairy farms, that the man didn't have time to go to work in the factory. But ours wasn't that big."

"They Worked Together Pretty Good"

Many native-born families also combined wage-earning with farming. When Penny Dumont Hall was growing up, her mother was the mainstay on the farm while her father worked intermittently in the factory. "He went to work at E-J and tried to farm at the same time," doing chores in the morning and evening before and after work. Penny's elder brother Frank, whom I interviewed along with her, recalled that their father "would work for two years, or four years, get tired of it and go home on the farm." But the family had only a small plot of land, so he always had to return to the factory. Although farming was his first love, the family needed his wages to get along.

The Dumonts had grown up in the same open-country neighborhood, attending the district school and Methodist church together. After they married in 1913, the couple lived first with Penny's mother's parents and then in a rented house. They had five children: Frank in 1914, a daughter in 1916, Penny in 1918, and two sons in the mid-1920s. By the time Penny was born, they were living on a 21-acre farm that had belonged to her paternal grandfather and renting pastureland from neighbors. Like most families, they had a big garden, and Penny's mother canned vegetables, fruit, and meat. Their major income-producing operations were dairying and poultry raising.

Frank recalled that his mother milked regularly when the children were young. They started off with only five cows, but eventually they built up a herd of ten. Each member of the family who was old enough would milk two cows. Frank started milking when he was around nine; so did his sisters and younger brothers. "I don't think my mother milked after I got started," Frank said. "When the boys got old enough, she didn't milk at all." But Penny explained that their mother stepped back in whenever her father and brothers were away from the farm. "My father worked with another farmer thrashing; this other farmer owned a thrashing machine and my father had a tractor, and they went around from farm to farm." The men in the neighborhood also filled silo cooperatively, which took them away for many successive days. When her father was gone, "my mother would help do chores." Before they had enough cows to ship fluid milk, their mother made butter. Penny and the other children used to help with the churning "because we liked to."

Their mother was "quite involved with the poultry business," which the family conducted on a fairly large scale. According to Frank, his father and the boys routinely took care of the hen house, fed and watered the hens, and gathered the eggs. But when they were preoccupied with seasonal tasks or working off the farm, his mother would take over. Penny recalled that "When us children got big enough, we did the egg-gathering, and helped her wash them, too." Their mother always "took care of packing the eggs." The family kept a laying flock year-round. Each spring, they would start

1,600 chicks, sell 800 roosters as broilers, and keep 800 pullets. Once a week, their parents took several cases of eggs to stores in town.

As Penny and Frank described their own and their parents' work, they spoke mostly about what "the farm needed." Penny, aware that she did not do much of what was conventionally regarded as women's work, spoke about her personal preference for outdoor tasks. "As a rule," she said, "the boys helped in the barn and the girls in the house." But her family was different because the farm needed everyone. And Penny herself was different, especially from her older sister. Penny spent a lot of time in the barn and fields. "I did, more so than my sister, I think because I wanted to get out of housework more than anything else. I liked the animals. I didn't like chickens, but I liked the cows, and I liked to get up early in the morning and go in the woods and get them and help out with the milking." "And it did come in handy" for Penny to do chores; "as the boys got older" they were more often away working elsewhere. Penny "didn't exactly *dis*like housework," but "my sister was older than me, and she was taught to do the baking and things like that, and I think that was the more rewarding or pleasant job than doing the dishes or something. I think housework is a monotonous job; you do it over and over and over." The only thing about farming that Penny did not enjoy was being limited by the milking schedule: "We disliked it when we went someplace and had to hurry home to do chores."

Not only were the Dumonts trying to build up their dairy herd, which meant enlarging and improving their barn, but their house burned when Penny was eight. That was "quite a hardship." But "all the neighbors chipped in and helped rebuild the house" while they lived in the granary. Fortunately, it was summertime. Frank recalled that "we had a picnic. Probably my mother didn't have a picnic, but us kids, we thought it was great." Penny knew they were poor, even in comparison to other families who lived nearby, and she heard talk of hard times and the Depression. She was particularly sensitive about not being able to get new shoes until they sold the broilers in the spring.

Penny and Frank agreed that their parents made decisions together, just as they worked together. The fact that their father earned cash wages at the factory and their mother did much of the income-producing labor on the land made no difference; they were equally committed to the farm. Penny had the impression that her father would listen to her mother's opinions, although they seldom disagreed. When, in a separate interview, I asked Frank who made the decisions in the family, he replied that "I'm sure my Dad talked them over with my mother. I suppose the final decision was quite often left up to him, because I'm sure it was mostly the men's, back in those days." But his mother knew everything about the farm operation, and "I think she could have changed his mind if she wanted to." Penny was most impressed by her mother's commitment to farming. She never complained that they had no inside plumbing and that she had to do the laundry on the back porch, for example. "I know there was times we didn't have things for the house, or didn't have much money to buy things. It seemed like there was always something needed on the farm. I guess that's true of every farm"; the barn came before the house "because that's where the money comes from." Penny said that her mother wanted to put whatever resources they had into the farm; after all, it was her workplace and she knew exactly what the family's financial situation was. As Penny saw it, her parents' division of on- and off-farm labor was not unusual. Although their contributions to the family economy differed, "they worked together pretty good."

"They Were Brought Up, the Man Is the Boss"

Penny Dumont Hall and Vera Socher Malone lived on adjacent farms near Bradley Creek in the town of Maine. Penny introduced me to her childhood friend after she had compared her family's farm with those of their Slovak neighbors who had recently moved to the countryside. In their joint interview, Vera told the story of her family, and she and Penny compared their experiences as newcomer and native.

When I first interviewed Penny Dumont Hall, she remembered that a number of immigrant families had moved onto nearby farms during the 1920s and 1930s. "We kids were always together." In Penny's experience, the children of immigrants and those of native-born families generally "got along well"; attending a one-room school broke down the barriers between strangers. The men often changed work, although the women "didn't neighbor back and forth" as much as the men did. Penny had the impression that women in immigrant families worked very hard. They did so much work outdoors in part because they were poor and had no machinery. When, as in Penny's own family, the father and older children held off-farm jobs, the mother and younger children had to do all the farm work. But, she thought, it was also a matter of custom. "I can remember Vera's mother, she came after their cow and milked it, nights; I don't know as her father ever milked." In her own family, by contrast, everybody helped with the milking. Penny thought that "in the Slovak families, the European families, the husband was the boss." She connected the women's burdens with the men's authority. For example, she was shocked that when the men came home from work the family seldom sat down to supper together; instead, the women served the men.

To a remarkable degree, Vera Socher Malone shared her friend Penny's critical view of family relationships in which men exercised authority as fathers and as husbands. That one family was native-born and the other immigrant made little difference in these women's opinions about gender and intergenerational relationships. Some of the resentment they expressed was directed toward the greater freedom accorded to sons than daughters, but some was expressed toward husbands who seemed to assert power over their hard-working wives. When I interviewed Penny together with Vera, both assumed that husbands and wives may have conflicting interests and that cooperation in marriage was an achievement, not something that women could take for granted.

The Sochers, who identified themselves as Slovak, were both born in a rural village in the Austro-Hungarian Empire. As a youth, Vera's father had served in the military and lived in Vienna, while her mother went to work in a sugar beet factory in Bratislava. When they met again, they married and had four children. Her father came to the United States in 1911, when he was twenty-nine. A Slovak travel agent directed him to Endicott, where he found work in construction. Then he got a job in the Endicott Johnson tannery, which offered more stable, year-round employment, so he sent for his wife and family. Three more children were born here. The Sochers bought a large house in Endicott where her mother kept as many as twenty Slovak boarders, washing, cooking, and cleaning for them all. Her earnings paid for the house.

In 1914 or 1915, the Sochers bought a 100-acre farm on Pitkin Hill. Vera thought that they paid half the money down, using the proceeds from the sale of their house, and took out a mortgage for the rest. Her father quit his job at E-J and dedicated his labor to the farm. They had dairy cattle and grew large quantities of potatoes and vegetables for sale. "My mother had a horse and buggy. She got up at 4 a.m. and she'd deliver milk, cottage cheese, butter, and eggs" to regular customers in Endicott. Their produce brought such good prices during the war that they were able to pay off the mortgage. Vera was born on the Pitkin Hill farm in 1922. But soon declining prices for farm produce forced Vera's father to return to Endicott Johnson, and the family moved from Pitkin Hill to a smaller farm near her aunt and uncle's farm on Bradley Creek Road. On 44 acres, which they owned free and clear, they kept just enough livestock to supply the family, raised hay, and grew potatoes for sale. Her parents loved living in the county, Vera said; after all, "they were peasants" in Europe.

Her mother took an active part in farm management, doing all kinds of work outdoors as well as in. She "did everything from fence posts to the dairy, between babies. She'd run in the house and get a meal, feed the baby, and back to the fields." After her father returned to the tannery, Vera's mother did almost all the farm work. Vera recalled that, when their husbands were at E-J, her mother and her mother's sister "pitched the hay all the way down to the barn forkful by forkful" because they had no machinery.

Vera's mother occasionally reflected ruefully that "they came here and had it harder than there." "She said herself sometimes, at least she had her mother in Europe; here she had just her sister and her own family." Not only were her hopes for a less laborious life disappointed, but she felt more isolated. She found the cultural adjustment especially difficult. Vera's father had applied for citizenship right away and attended English language classes before they came, but her mother picked up English primarily from the children. The older ones had to learn English at school; their mother would not even attend PTA meetings. The children often conducted the family's business dealings with Americans.

Although Vera grew up as a native English speaker and learned American ways, she felt the strong impress of her Slovak background. "To tell you the truth," she confessed, "we were always a little ashamed, because, we always felt, we were foreigners....We always felt a little inferior to the old settlers. When we were growing up, we always felt that we didn't belong."

The area of Pitkin Hill where the family first lived became known as "little Europe." The immigrant families who bought farms there included the Ivans and the Vaniks, who were also Slovak; the Dworniks, who were Ukrainian; the Vargas, who were from Byelorussia; and the Maggios, who were Italian. Most of the people living near them on Bradley Creek, by contrast, were native-born Americans. Their immediate neighbors, particularly the Dumont family, "were wonderful," Vera enthused. They helped the family with farm work and translation, and the children were the best of friends. But Vera's mother did not associate with the other women in the neighborhood. "She was more or less tied up with her Catholic Church and their own Slovak people" in Endicott. "We were always on good terms with the Dumonts and the neighbors right around, but" when her mother had to associate with Americans she "was always in the background."

Vera felt fortunate to be able to graduate from high school. "The older children had it rough" because they had to go to work earlier than she did. By the time she was fourteen or fifteen, they already had jobs. The older children's earnings paid for things for the house, including a washing machine and the installation of electricity. Each week they gave all their wages to their mother, who gave them back $2 for bus fare and spending money. "In the old country, until they married, the father and mother were the head of the family."

She described her own family as male-dominated. "Father always had the say-so." Immigrant husbands expected their wives to work hard and did not treat them with the deference that American men showed, she said. "They were brought up, the man is the boss, the father is the head of the house." This dictum applied to the children as long as they lived at home. "They always favored sons more than daughters, *always,*" Vera exclaimed, her indignation audible even after so many years. "Sons were *worth* more than daughters." Vera felt that most immigrant families "were very strict with the girls." When she tried to complain to her father about this disparity, she got no response—not even a reason for his refusal. "He was good to his children, but...*mind*. No explaining. Nobody talked back to him. You never questioned his word; what he said once was law." Vera felt that her mother was "more sympathetic and kind-hearted. Mother you could cling to; she was comfort." She tried to get her mother to speak up and support letting her do what she wanted but, Vera said in a resigned tone, "she never did."

Penny Dumont Hall said that her own father was even stricter than Vera's. He forbade the girls to take walks on Sunday, for example, because Methodists disapproved of such frivolous pursuits on the Sabbath. Penny emphasized that her mother was not subjugated to her father as Vera's mother seemed to be, and her father paid more attention to her mother's opinions. But when Penny was interviewed with her female friend, rather than with her brother, she spoke more critically about the tendencies toward male domination she saw in many rural families. These lifelong friends agreed that men held more power than women: fathers wielded authority over children, and sons were privileged over daughters. As teenagers, Penny and Vera swore that they would never marry men like their fathers; they wanted husbands who would show more understanding of their children. To a substantial degree, Penny and Vera identified themselves not as a native and a newcomer but as young women who were seeking the freedom to define their own lives against the restrictions that their fathers imposed. Still, they agreed that immigrant husbands held more authority over their wives than American husbands were able to exercise. In this respect, Vera's friendship with Penny reinforced her resolution to have a more equal and reciprocal relationship with the man she married.

"She Often Said What a Hard Life It Was"

Following the lines of family connection, I interviewed Penny Dumont Hall's lifelong friends Naomi Bowen, who married Penny's brother Frank, and Ivy Bowen, Naomi's elder sister. According to the Bowen sisters, their father, mother, and siblings shared and divided farm work flexibly, depending on the number of hands required in the barn and the number of people available to work. Yet authority relations in their immigrant family were similar to those in the Socher family; their father was the boss, and their mother went along with what he wanted. She seemed to find farm life less fulfilling than Vera's and Penny's mothers did.

The Bowen family moved to a farm located on the road from Maine to Newark Valley in 1927, when Ivy was fourteen, Naomi was eleven, and their father was fifty-two. Both their parents were Welsh; their father had been a coal miner in Wales and in Pennsylvania. The family had moved to Endicott so her father and older siblings could work in the boot and shoe factory. Eight younger children remained at home when the Bowens traded their house in Endicott for a mortgaged farm. The impetus for this move came from their father. Naomi recalled that "Dad was working at E-J in the tannery, and he had a problem with the boss or something and quit, so there wasn't much alternative. He'd always wanted to farm, so that was the way we had to go." Her mother "would follow my father any place," Ivy affirmed. When I inquired how their mother liked living on the farm, Ivy asked Naomi, "She never said she didn't like it, did she?" "No, but she often said what a hard life it was," Naomi replied. Their mother came from a coal mining family and had always been poor, "so she didn't have much to compare it with." The main difference between the city and the country was the lack of basic utilities. Ivy recalled their mother saying that when they traded the house for a farm, "We got lamps, they got electric lights."

But, Ivy declared, "I didn't give it a thought. I loved being free on the farm. Of course, we had to work, but that was nothing." Nobody in the Bowen family knew anything about farming, so they learned through "trial and error." Their older sister lived on a farm just above theirs and her husband had grown up on a farm, so their father "went and asked him questions." Ivy never went back to school after that. Instead, she "milked the cows, pitched the hay, took hold of the horse and led him along to cultivate" the fields. She got a job at Endicott Johnson when she was sixteen, but she still did chores every morning before leaving for work.

Slowly the Bowens enlarged their dairy herd and paid off their debts. It took three people to milk a dozen or more cows. Naomi recalled that "my mother milked, and as each of us kids got old enough we milked. We all went to the barn, girls and boys." Ivy had the impression that their mother did not milk so regularly after the children were old enough to do so. Naomi agreed that when she got old enough to milk alongside their father and her older sister or brother, their mother "wasn't needed in the barn." Later on, when the older children stayed on at work, she milked in the afternoons, at least until the youngest girl was old enough to help.

Naomi particularly recalled their mother wearing a dress and apron in the barn; "I thought it was so un-handy to milk with a dress on." When she started milking, Naomi made what she called pajamas out of feed bags to wear to the barn; the wide legs enabled her to move more freely than her mother could. Later the younger girls wore boys' overalls to do chores. Ivy recounted an embarrassing story about being caught in the barn by her supervisors from the factory. "One day he and my forelady came out to visit, when I didn't expect it. And I was sitting there milking a cow." She laughed and continued, "My mother sent them down to the barn, and there I was milking. I could have fell through the floor!" "At that time, you were embarrassed to be a farmer" when you were in Endicott at school or at work, Naomi commented. Ivy continued, "I didn't want my boss to come out and see me! Of course, you're rather ungainly sitting and milking a cow, you know. You have one foot across the drop, and the pail's in here"—she gestured between her legs—"and, oh boy! I wanted him to think I was a *lady,*" she concluded emphatically. The sisters debated whether a lady could be a farmer. They agreed that was possible, but they had not thought so when they first moved out from the city.

The girls and their mother separated the milk and made butter, butter-milk, and cottage cheese to sell. Ivy recalled turning the barrel churn for her mother. Once she spilled the cream on the floor because she had failed to tighten the lid after her mother checked its progress. "I felt terrible, because we needed the money. I don't think my mother ever bawled me out for that. I never did it again; I watched very carefully." Later the family started selling the cream through the Dairymen's League and then shifted to shipping fluid milk. Eventually they had a water-cooled holding tank in the barn, but they never had milking machines.

Ivy and Naomi recalled besting their father when the whole family was haying. Ivy began, "My father always threw the cocks up onto the wagon, and I placed them around the wagon; that was my job." Laughing, she continued, "When my father tried it, he tipped the wagon over." Naomi took up the story: "That's what all of us girls used to do, because he said we couldn't pitch it up on. So we traded with him once, and we could pitch it up on, but he didn't get it placed good." Naomi did not recall her mother helping to get in the hay, but all the women shocked oats and dropped and dug potatoes. Naomi thought that perhaps the women in her family did so much field work because there were few men in the household after her older brother left home. "He and my father had some differences, so he left." The farm work had to be done by her father and her younger brother, her mother, her two sisters, and herself.

When I interviewed Naomi Bowen Dumont with her husband, Frank, she said, "I'm sure my Dad made the decisions of when to plant and how much to plant, but if it was a big decision about buying more things, they talked it over too, if it meant an investment of money." In the sisters' joint interview, however, Naomi and Ivy agreed that their father was the one who exercised authority in the Bowen family. As Naomi put it, "He was the, kind of the boss." "Yes," Ivy assented. "He never spanked us, but he was the boss, and we knew it." For example, when the girls wanted to attend a church social they'd ask their mother first, and she'd usually ask their father—although their parents never discussed such matters in front of them. "We never asked our father for anything. It was a funny relationship, wasn't it?" Ivy mused. When they lived in Endicott and she needed a book for school, she never asked her father for the money. "So when I didn't have it," the teacher "sent me off the principal. I wouldn't tell her why I didn't have it, I wouldn't talk; I just cried. So that's the way it was." Perhaps to shake off the humiliation that their poverty and paralysis entailed, the sisters laughed uncomfortably. When I asked them why they were so reluctant to ask their father, Ivy replied, "I don't know why. We were just brought up that he was the boss, he was the head of the house." "We knew we didn't have the money," Naomi added.

Ivy and Naomi's mother had no way of earning money that she could control. As Naomi put it, "She never had anything separate from his; it was

all his." "She got what she needed," Ivy interjected; "my father was gener-
ous enough with her, as far as that goes. I never heard them squabble over
money." When Ivy went to work at E-J, she gave her mother the $20 she
earned every week and got back enough for bus fare and lunches. Later
Naomi made $25 a week at IBM and gave her mother $10, as well as buying
groceries. "It was just the normal thing to do," Ivy explained. "When your
parents needed money, you went to work and you helped." "All our sisters
did that," Naomi concurred.

The sisters thought that their mother and father shared the view that
the barn came before the house. "It had to be, because you had to have
those things to make a living," Naomi explained. The Bowens had few
household conveniences. Ivy recalled that when they bought the farm, "we
sat in the kitchen and the wind was blowing in the window and rattling the
curtain." Laughing as if to dispel her memories of discomfort, she added, "I
used to put newspapers on the bed between the blankets to keep us warm."
The Bowens got electricity in 1937, when the rural electrification program
reached their isolated neighborhood. As long as Naomi lived at home, they
never had inside plumbing or a washing machine; all they had was a well
with a pump, a hot water reservoir on the back of the stove, and a sink with
a drain. Fixing up the house was not a priority; money spent on household
conveniences was not an investment that would pay off.

"I Drove a Tractor and Worked out in the Fields"

In their own married life, Naomi Bowen and Frank Dumont had the same
priorities. They had no electricity in their house when they moved onto the
farm in 1939, and they did not install a bathroom until 1948 or central
heat until 1953. Naomi explained that modernizing their house was delayed
"because our barn burned down, and we had to build this barn in 1945, and
that kinda knocked out plans for the house."

Frank and Naomi started going together in high school. They married in
1937, when he was twenty-three and she was twenty. For the first two years,
they lived with his parents while he worked at E-J. Naomi had to leave her
job because IBM did not employ married women during the Depression, so
she helped Frank's mother with the eggs. "It wasn't an easy two years, but
we done all right," Frank remarked. Naomi explained, "He was the oldest
of six children and they were all at home, so I had to get along with them."
His sisters were close to her age: "That was the hardest part, getting along
with them girls." When she married, Naomi acquired not just a husband but
his whole family—"or, rather, they acquired me."

The Dumonts planned to buy their own farm from the time they were
married. When she was younger, Naomi explained, "I said I was never going
to marry a farmer, because my father did everything the hard way and I
thought it was too hard a life. But then I fell in love with Frank and I knew

he was going to be a farmer, so I did it anyway." Frank saved $1,000 out of his wages and she saved $500 from hers; they planned to use his savings for the down payment on the land and hers to furnish their house. In the end, however, they spent all their joint savings on the farm. It adjoined Frank's grandparents' farm and was close to his parents'. He borrowed his father's machinery for a number of years while he and Naomi built up their dairy herd.

For six years, until 1945, Frank kept his job at Endicott Johnson. He affirmed that the poultry business "paid for the farm," in much the same way it had done for his parents. They raised capons and sold them in the fall for around $2,000, which they used to pay the mortgage. By raising calves, they gradually enlarged their dairy herd. Frank and Naomi did the milking together. Initially she separated the milk and churned butter in the kitchen, selling it to customers who came by their farm, which was located on a main road. When they shifted to selling milk, she was "relieved. It's no fun to make butter; it's a lot of hard work." They had neither electricity nor a refrigerator at that time, so it was difficult to keep the cream at just the right temperature for churning. When they were milking 14 cows, Frank finally quit working in the factory. But then their barn burned. "I thought about going right back to work," Frank said. But "the neighbors were real good. They took up a collection and gave us about $900. A lot of people helped work on the barn," too. It was finished on the day their fourth child was born.

Naomi participated in most of the farm work, even when the children were small. "I liked it....I'd say it was hard work, but I guess you just didn't think, you did it." She felt like an equal partner in the farm enterprise. "I drove a tractor and worked out in the fields. I didn't plow or cut hay, but I would rake, and drive to take the hay in. I'd drive the loader, and get the hay off." She went to the barn twice a day. "I'd throw the hay down, and feed the ensilage." Naomi did more farm work than her mother had done because her mother "was older when she began and she had children who were old enough to go right out the barn, which I didn't. After our children were old enough to go down, then I didn't do so much either."

Naomi's labor also became less crucial because they mechanized milking. When the Dumonts rebuilt their barn in 1945 they installed a bulk tank; later they got a pipeline milking system. Naomi never worked much in the barn after that. She said that the new system was too high up for her to reach, but Frank protested that if her labor had been needed they could have gotten her a stool. Frank could milk 50 cows by himself morning and evening.

None of Naomi and Frank's children wanted to go into farming, even though they enlarged the farm from 72 to 150 acres and built up a profitable dairy enterprise. All four sons became engineers. Frank commented that "it might have been nice" if one of the boys had chosen to farm, but "it would have caused some problems, too, since the boys' wives aren't farmers." Naomi "wasn't anxious to" have any of the children succeed them "because at the time they would be coming into the farm we would be getting closer to retirement age, and it would mean that you'd have to go in heavier [rather] than getting out." They had seen other families who incurred greater debts and worked even harder than before just to pass on the farm to a child. Naomi and Frank never hired much labor, but as they got older they hired a trained herdsperson. She was very good with the cattle, but she was not a relative and had no way of buying them out. The Dumonts had sold their cows several years before I interviewed them. They rented the land to a neighboring farmer, and Frank worked for him quite a bit. Both he and Naomi took pride in the productivity of the land that they had developed together over a lifetime of labor.

"That Was My Love, the Outside, to Work"

The most eloquent statement I have heard about a Nanticoke Valley woman's relationship to the land she farmed comes from a Ukrainian immigrant. Nadya Stastyshyn, born into a poor peasant family in 1890, came to the United States at the age of fifteen. After toiling in textile and shoe factories in New England and Binghamton, Nadya and her husband, Simon Fenson, bought a farm in the Nanticoke Valley. Nearly sixty years later, in 1978, she was interviewed by a woman whose ancestors were from the same region.[2] In language that bears audible traces of her origin but is extraordinarily powerful, she articulated what the land meant to her.

Nadya's oral autobiography demonstrates the extent to which a woman's sense of agency can develop through her work on the land. Living on the farm became central to Nadya's identity. Making things grow was a source of joy, and making money was a source of self-respect. Indeed, moving to the farm transformed her sense of self. In the earlier sections of her autobiography, Nadya described herself as controlled by necessity and the will of others. When she arrived in Rhode Island, "they took me to the factory, so I work in the cotton mill"; when she first came to New York, "my husband bought a house, so I had to stay at home." As she told about the years that followed the move to the farm, however, Nadya positioned herself as an active agent in her own life story. She described making decisions and controlling her work. Her relationship with her husband became more of a partnership as she—and Simon—discovered the value of her labor and her business acumen.

After Nadya married Simon Fenson, they continued working in textile mills in Massachusetts, but "jobs were scarce." She had an uncle living in Binghamton, and he "told me to come to Binghamton, as here we have the Endicott Johnson factories and lots of people are working here and getting overtime pay." They came in 1917. Three years later, "we traded the house for a farm."

From 1920 we came on this farm. We were poor; it was a hard life; we had a dilapidated farm. Ah, little by little...My husband was working in the factory, but there wasn't much work so he had to quit. We decided we would farm. He didn't know anything about farming and I didn't understand farming, but we were young, so we figured we would get along. I was in town occasionally and saw, every once in a while, a farmer would come into town with his horse and wagon and bring in different things to sell. I was thinking, I too will try, take the horse and wagon downtown and sell something. We owned three cows, so I made cheese and butter and took it with me. Many customers bought these from me. Ah, people bought from me but few paid for it. Times were bad. We were poor, no money coming in from anywhere.

Once, as I was out delivering butter and cheese,...a[n] insurance man named Bay...stopped me and asked, "Housewife, how is business?" I replied: "Business is good, people are buying, but nobody is paying anything....They buy my butter and cheese but tell me, 'I will pay next week as now I have the electrical bill,' all kind of excuses not to pay me." He told me, "Stish, why do you go and knock on doors to sell your wares? Why don't you go to the Johnson City Public Market?" "Bay, I have never heard of that market"; I never had. He said, "Listen, I'll make a map for you," so he proceeded to take out his little book and on a piece of paper he drew a map with directions to get to the market. He said at the market people come to you, buy your wares, and pay CASH. I thanked him.

That night, I was telling my husband, the next week I will not to go my cheese and butter customers as Mr. Bay told me to go to the Johnson City Public Market. My husband replied, "Oh, you are foolish to listen to people. You got good customers. Take care of them!" I said, "What good are they? I never get paid for my labor. That's the business!"

Later in the interview, she described their disagreement in greater detail:

I say to my man, I say, the next week I go to the market because I hear Mr. Bay, told me market people pay cash. He said, "Don't fool yourself; you got steady customers." Yeah, those steady customers take everything, but nobody pay me a nickel. Well, I went like Mr. Bay told me, my man was so mad he no want to put, he no want to hook the horses to wagon. He put harness on— horse was big one, I was short, I can't put harness on. He don't want to hook horse to wagon because he said I should not leave my customers. But after that he don't say nothing. He find out I don't do nothing wrong.

"Well, it just happened I was so fortunate to have a large crop of peas, such beautiful peas, that I picked two bushels, took along cheese, butter, etc." She worried that "I could not speak much of English." When she arrived and saw all the people with horses and wagons, "I was a bit flustered....But here came Mr. Patterson, manager of market. He said, 'Welcome, welcome, new producer!'"

Then he glanced at my wares and said, "Good! Good!" He took some tags from his pocket, saw my peas and came around with a quart basket, and says, "This basket is 18 cents a basket or 2 for 35 cents." Alright, when I started selling—by gosh! I couldn't believe the people at my stall, buying my wares. I couldn't keep up with my customers. I didn't have to have bags, as the people paid for my peas and told me to dump my peas into a basket they carried on their arm. I sold everything. The manager approached me and said, "Come back next week." "Good," I said, "I'll be back." I came back every week, brought anything I had to sell, and everything went I received CASH! I looked around and saw what people were selling to give me idea. I just couldn't get over it, anything I brought in it was sold for cash—nothing trust!

Mr. Patterson "helped me, everything. He said, 'When people buy from you, you have to be very nice to them.'...He teach me; I don't know nothing about nobody. He said, 'When anybody come to you, they buy from you, be very pleased. When you give them change back, say thank you—come again!' That's what I did."

That first day, she figured out how to placate her husband.

I got the cash, I got $16, see $16 I never see in years. I know my man he was worrying. He got horses; they need oats when you go on field work. We had no money to buy oats. When I got this money I go to farmer store and I stop and I ask the farmer stores man, I say, "Mr., I got one horse but want to buy a little grain for horses, anything I needed. Can that be too heavy for horse to go?" He said, "No." [Laughter] "How much do you want?" I say I want it, at least 2 bags; that's 200 [pounds] horse feed. "That's not heavy atall; he can take lots more." Then I take it, I take 2 bags horse feed, 1 bag cow feed, and I got myself pork loin, pay everything cash and come home. I still got $4 cash— beautiful! Beautiful! When I got home, my man he said, "What you got in those wagon?" I said, "Why don't you look over?" He see, boy, he grabbed the bag, he said he have to go on field, the horses don't have nothing to eat. Then afterwards, he don't say nothing to me. You wanna go, go; you wanna go, go.

Nadya began selling her vegetables at the market in 1921. She continued to learn from other vendors.

I raised vegetables, but vegetables very little profit; too much work and not much good of it. I had vegetables like carrots, onions, beets, etc. on a bench, but outside they wilted in the sun. One summer day, a lady of sixty-five or more who had a stall nearby came to chat with me. As we were talking, she pointed to my vegetables and said, "Lady, you had beautiful vegetables there but now they are wilted. People will not buy wilted vegetables." I answered, "What should I do?" "Forget the vegetables," she advised. "Why not raise flowers?" I replied, "Flowers! Who needs flowers?" I didn't believe her. She continued telling me that "anyone who owned a home and had a small plot of land in back they always grew vegetables for their own use, but flowers, they have no room to plant. If you sell flowers you will have a good business." I don't know how to go about this new venture. The lady proceeded to tell me that as soon as the snow melts in the spring and the ground is not too wet, plant some sweet peas and they will grow nicely. I did just that.

Again, her husband was skeptical:

I wanted to buy flower seeds. My husband laughed and said, "Foolish lady, who is going to eat your flowers?" I told him, "I'll try it." I spent one dollar on flower seeds. After a short time, I told my lady friend my sweet peas are growing so beautifully, such beautiful flowers, can't get over that. What shall I do with them? I don't know; I had at no time seen arrangements of flowers. She replied, "What? Get scissors, cut them, and make bouquets."...She was very obliging, showed me how to. Later she said, "Use your head, too," because if you make a beautiful bouquet you will have many sales, but if you just bunch them they will not be appealing and you will not have a sale. I tried the best I could.

So I had flowers, but it was a job, especially the cutting and making bouquets. Now I started to plant other flowers, even planted the flowers my customers requested. I tried. I noticed a seed catalogue, I ordered flower seeds. My customers were pleased. I had such a beautiful assortment of flowers; I had asters, zinnias, gladiolas. I sure found success with my flower sales. Many people did not believe me, that I could earn more money at the market than the lady who works in a factory. I found if you have the will, the strength, you can do very well.

When Nadya began selling in Johnson City, it was an open-air market. Later it was covered so it could operate year-round. It was no longer free; "we paid $12 a week, but it was worth every cent." She recalled an especially good year:

When I sold glads, that alone was good money, but instead of selling each flower I made bouquets for $1.00, medium sized 50 cents. People bought for cemeteries. I had beautiful flowers that I kept fresh in pails of water; people were standing in line to buy them. My flowers sold every time, but if I had a few left, I gave it to people for their church. They were a good profit for me. I remember 1939, it was a fantastic year, sold so many chickens, assorted flowers, glads. Gladiola sales alone, I made $135, was pretty good. I always kept track of my sales, always sold everything.

Nadya began raising poultry and selling dressed chickens. Her husband assisted her with this project.

There was a man from Owego who sold all dressed chickens. I decided, well, I want to sell dressed chickens, too. The next week I took with me crates of live chickens, 16 chickens in a crate. At the market, people went by, looked, said "nice chickens," "nice chickens." But were not buying my chickens, no. One of the following days, it was rainy weather and my husband could not work in the fields, told me he was going with me to the market, as company. He helped me bring out my wares plus the crate of live chickens. I am think-ing all the time, I have such beautiful chickens. How can I make that chicken dead? Nobody is buying my chickens. I am still thinking, thinking, and all of a sudden I got an idea.

While her husband was talking with several other men, she took two roost-ers and a hen and went to a friend's house nearby. Nadya asked her to put some water on the stove so she could pluck the chickens.

She asked, "Who will kill the chickens?" I replied, "I don't know." But we finally agreed that since both of us cannot do it, we asked the neighbor. She was obliging: "Get me the knife; I'll kill them." Finally the chickens were defeathered and I returned to the market. I saw my husband as I left him talking with the men. I asked, "Did you sell anything?" He replied, "No, no one stopped by." Now, I put my dressed chickens on the bench. Before long, a Slovak couple come by and stopped to look at the chickens and asked if they were fresh. I said, "They are shaking, are warm." The lady answered, "My God, they are warm." Man answered, "Oh, the sun warmed them." I convinced them my chickens were fresh so they bought the largest rooster for baking. I was beginning to worry if will sell all the chickens but as it was, another couple came along, another person came along, so within 30 minutes my dressed chickens were sold.

It was against the rules to sell meat in the open-air portion of the market, and another vendor complained to the manager, "saying, 'that lady is tak-ing my business away, get rid of her.' In all fairness, Mr. Patterson replied, 'Harry, you take care of yourself and she will take care of herself.'"

As time went by, my chicken business grew from 20–30 chickens every Sat-urday to 100–125 chickens. At holiday time I included ducks, geese. At times we had to buy chickens to keep up. Finally, I had such a business, my hus-band wanted a picking machine but it cost $300. He figured that if someone had one I'll have one. So with pieces of wood, metal, he made one of his own which worked beautifully. The chicken business was a lot of work and a lot of fun. So that was our life, 35 years at the market, from beginning to the last.

During these same years, Nadya and Simon were operating a dairy. "In the beginning, because we had only three cows," they sold butter and cheese at the market. "Later, little by little we bought more cows. We had milk, the milkman picked it up." They bought a milking machine for $350, which meant that her husband could handle the milking by himself. "It was wonderful; didn't need to milk by hand."

Nadya spoke proudly about how they developed the farm. They rebuilt their dairy barn and built a henhouse, brooding house, and incubator for the poultry. They did not improve their own house, even though it had few conveniences. "It was difficult living here" at first. "In this house there was no, no *nothing* when we bought the farm. We had to go outdoors and pump our water. See, in the house we did not have running water, only a well. But we lived and everything was alright." If their house had not burned down in 1940, they would have continued to make do.

Nadya valued both the natural environment on the farm and the working relationship she developed with her husband. Her recollections of working with Simon in the winter woods epitomize what she enjoyed most.

> Sometimes, during the winter weather when snow was so high, you could not see very far, my husband would start off for the woods and say, "Do you think you could leave the children alone and go to the woods and help me?" We had about 35 acres of woods. You could get a lot of lumber out of them, but some of the trees needed a cross-cut saw so I would go with my husband and help to hold the saw as he cut the lumber. While my husband is cutting the wood I'm piling the wood, helping him in any way I could. I'd put on boots and go in the deep snows. When you are out in the deep snow, you don't realize that it's cold, because you are working.

Nadya and her husband had to stop farming in 1953, when the public market closed. "We sold our farm that year as raising vegetables, chickens, glads were not necessary if I did not have a market to sell them at." Ill health had made it impossible for her and her husband to work outdoors. "We sold everything with the farm but two acres, where we built a small house. The land is idle. I can't work it anymore. There it lies dormant. When I was able to work I had everything, raspberries, vegetables, even flowers. Now, I look at the idle land and it hurts; it pains me to know I have the land but can't work." Nadya identified with the farm so intimately that she thought of both the land and herself as "idle" in old age. The pain she felt came as much from seeing the land grow up to brush as from the arthritis that prevented her from cultivating it.

At the end of her interview, Nadya affirmed that

> Even with all of life's tribulations, if I had to relive my life, honest to God, I would relive it in the same way.... Only give me back my strength. I just love to work in the fields, in the garden. That was my love, the outside, to work. I'm not like the ladies who get together for a coffee klatch or bridge; theirs is a different life. I loved the outdoors.
>
> When we were younger, we were able to work the fields, the garden, milk cows, tend to the chickens. I liked planting new things. When I saw different

interesting plants or vegetables, I always wanted to have them, too....I had everything! I loved to work.

Through her work on the land and her sales in the market, Nadya developed a strong sense of personal agency, and her husband became a more cooperative farm partner.

8

Negotiating Working Relationships

What made the most difference in the tenor of rural women's life stories was the degree of agency they felt able to exercise over the major decisions that shaped their lives. The women I interviewed had not grown up expecting to conduct their lives as autonomous individuals in the same sense that their daughters and, especially, their granddaughters did. As they looked back on their experiences and articulated their way of thinking, they described themselves in relation to their families and kin and as embedded in a network of connections that extended through their neighborhood and community. Like most farm people, they expected to have to contend with many circumstances over which they had no control. Growing up female, most did not know enough about their bodies to be able to decide whether they got pregnant and how many children they bore. But farm work was a more gender-neutral domain in which they could demonstrate competence and expect appreciation, or at least respect.

Most of the women who remained in or moved to the countryside sought not autonomy but mutuality; they deliberately sustained relationships with others that were characterized by flexibility and reciprocity. Their relationships with the men in their lives—fathers, brothers, husbands, and sons—were shaped by the structures of male dominance that pervaded American society, but these women were aware of the importance of their own labor and commitment to the family farm in counteracting the gender divisions and hierarchies of power that seemed to prevail outside their rural culture. They negotiated with their husbands over what work they would do and spoke up when decisions were being made about the farm. Those who founded farms with their husbands, rather than marrying inheriting sons, were able to shape their own work routines to a remarkable degree and had a strong voice in family councils. Strikingly, those who contributed money they had saved from their earnings to the down payment for the farm they established with their husband were more likely to have their name on the deed to the land than those who did not. Yet even women who married men who farmed in partnership with their father and/or brothers found ways to create marital partnerships characterized by mutuality rather than marginality.

Some women who grew up in circumstances of abuse or neglect and who later found themselves in marriages in which they felt disrespected became convinced that they were unable to exercise even a minimum of control over their own lives. Their stories are much harder to recover than those of women who felt more fulfilled by their lives, at least in retrospect.[1] Those who were in a position they found utterly intolerable fled their families and the countryside entirely. Yet some of their stories can be reconstructed from the memories of those they left behind. The circumstances that such women found most oppressive and the experiences that finally enabled them to take action are especially revealing.

The Revolt of Maggie

Carolyn Lane Charleroy struggled to understand the breakdown of her parents' marriage. In 1946, when she was twelve years old, her mother, Maggie Saunders Lane, went off with another man, leaving a new baby and five older children behind. Carolyn's father, Sampson Harwood Lane, tried to raise the children alone. At thirteen, Carolyn went to live with and work for a middle-class family in the village, and at seventeen, she made a respectable marriage. Much later, the siblings met their mother again. But Carolyn still felt confused about what had happened during her childhood. Looking back from the position of a wife and mother, she wondered what could possibly have led her mother to take such a desperate step. She and I also discussed these troubling events with her mother's younger brother, Wesley Saunders, who took Maggie's part, and with Carolyn's oldest sister, Clarice, who defended their father against their uncle's criticism. All these conversations centered on two closely connected issues: her mother's experiences of childbearing and working on and off the farm, and the sources of conflict between her parents.

All agreed that the Lane family's troubles had begun in the previous generation, with the dramatic dissolution of Sampson Harwood Lane's parents' marriage. The rupture in their relationship, which occurred when he was just five, was a scandal in the community, and their separate departures left four young children behind. Sampson Harwood (whom some called by his first name and others by his middle name) never had a stable home when he was growing up. Focusing on the positive and emphasizing the presence of his extended family, Carolyn said, "His grandmother brought him up, and then when she couldn't take care of him he'd go to live with his uncle, or work around." Still, he suffered from extreme poverty. As an adult, he recounted numerous instances in which his feckless relatives took advantage of his labor, which kept him from getting ahead financially. Above all, he suffered from the stigma that his parents' conduct had brought on the family: "My father said that he worked hard to keep his name clear," and he was determined that "it was going to stay that way." When his wife left their family and sullied his own reputation, Sampson was especially bitter.

Eventually he was able to buy his grandmother's farm, which had been in the family for generations. The large, four-square house, which stood at a crossroads on top of a hill, was built in the 1840s with timber cut on the property. Wesley Saunders explained how Harwood came to buy the place. "His grandmother," who had remarried and moved to town, "told him after he growed up that if he would come down there and stay to home and save up his money, that she would let him board there for nothing. So he done it, and he saved up $2,400. He come up there on that farm.…and he eventually bought that place for $10 a month. I don't think he had it paid for when him and Maggie got married." She had "work[ed] in the shoe factory. Maggie must of helped pay for the place as she was on the deed."

Sampson Harwood was twenty-four when he married in 1924. His wife was twenty. Maggie Saunders had grown up nearby as the oldest of fourteen children. Carolyn explained that as a girl Maggie "was like a mother to all of her brothers and sisters, because her mother was always busy having another baby." But Maggie preferred working in the barn and the fields to doing housework. The Saunders family rented, so they moved around a lot. For a number of years, they lived on a large dairy farm on the Newark Valley Road, where she met her husband.

A Younger Daughter's Unsettling Memories

On the Lanes' farm, according to Carolyn, Maggie "always helped with the milking. She'd go down to the barn every night and every morning. Usually my father would go down and get started, and then she'd come down and milk her five or six cows." The family generally had about twenty cows, with between thirteen and fifteen milking at any one time. They sold fluid milk, taking it first to Maine village and then to Newark Valley. Carolyn's father had a truck and "used to haul milk from the neighbors to there in the morning, and then come home and farm in the afternoon." The dairy was the family's main source of cash income.

The poultry flock also brought in money. The Lanes purchased White Leghorn peeps by mail order every spring. Maggie and the children were mainly responsible for tending the chickens and gathering the eggs. First

Sampson "had an egg route down in Endicott. He would take eggs and chickens and vegetables from the garden, if we had an excess." Later "he had a stand in the Endicott Public Market," where he went by himself every Saturday.

As the children got old enough, they all helped on the farm. "Now, my two oldest sisters, I don't think they went to the barn as much as I did. There were no boys when I was young. When my father would go hunting and trapping, I would go with him, sort of like a son." "We always helped my father just like we were boys." Like her mother, Carolyn would try to do whatever was needed, such as driving the tractor to pull the truck out of a ditch.

"My mother was always working outside," Carolyn recalled. Maggie raked and bunched hay. "We never had a hay loader and we never had a baler or anything like that; we always had a dump rake....We'd bunch it in the morning and put it in the barn in the afternoon. We'd go up" to the house at noon "and eat something, and then go back and work in the hay-field. Then it was time to go get the cows, because we had the cows way over on the back pasture. We'd bring them over, and then when they [her parents] was milking we [the girls] used to get supper. Then after supper we'd all go out and work in the garden while my mother was canning."

"We always had a big garden." Carolyn's father plowed and did the horse cultivating, while the children weeded. "We didn't ever buy too many groceries. There was so little money. Usually what we ate came out of the garden." "If we ate beef," it was what they had raised, butchered, and canned. "You never killed a beef just because you wanted meat; you waited until a cow died, and then you put the meat up quick. My mother was really from the old school. They used everything." From their hogs, her mother "made lard, sausage, head cheese, and pickled pigs' feet; she also used the tongue, brains, heart, and liver."

"My mother always enjoyed working outdoors," Carolyn reiterated. "She would rather do that than work in the house, anytime. In fact, the house always showed it. But we were happy. Everybody's house was always the same, because all the mothers always helped outdoors. Except some that, like E-J workers who quit their jobs to come up and farm, those mothers always stayed in the house. I could never figure out why they were not helping, because my mother always did." On reflection, it was clear that these women had grown up in the city and lacked the skills that farming required. When Carolyn was young, "it always seemed funny that their mothers weren't helping, like our mothers always did." Her best friend's mother, who lived next door, "went right to the barn." "My mother plowed, she dragged and sowed. They always worked together, my Mom and Dad. I don't know why they didn't hit it off."

One of the main causes of her mother's desertion emerged as Carolyn told about her youngest sibling's birth. By that time, Maggie was working off the farm at Endicott Johnson. Carolyn's brother was around nine, so he was able to help their father when their mother was at work. As Carolyn told the story, leaving the fields for the factory was her mother's own decision. "During the war, E-J really needed help. My mother finally went [back] to work there.…She worked until the day before my sister was born, she had the baby, and then she never went back to work. She just wasn't happy after the last baby was born. She was depressed and different, and didn't care much about going out and helping" in the barn and the fields as she had always done before.

At that point, Carolyn broke off her narrative and expressed the puzzlement and distress she had felt as a child: "I don't know what happened, really. It was such a shock to all of us kids. One day we got up on a Sunday morning and my father said, 'Your mother's not going to be here any more.' And then after a while she came back for a couple of weeks. She got the house all cleaned up, and away she went again." Once Carolyn had recovered her composure, she continued, "After she had that baby, she was never really with it. I don't know if she ever talked about it, or if she just kept it to herself. But I think that when she went out to work, that's when she realized that there was more to life. And then she got pregnant again, which was" a surprise. "Probably she thought she was all done having children" because she was forty years old. "After she had the baby, everything seemed to go haywire in our family."

Carolyn's account made it apparent to me that Maggie had suffered from postpartum depression. The older girls had worried that their mother barely noticed the new baby. She had always nurtured the little ones and was especially excited when her son was born. This time she seemed strangely passive, almost paralyzed, and she cried a lot. Carolyn had never heard of this condition before I mentioned it to her, and she felt immensely relieved to learn that some women suffer from depression when their hormones are readjusting after childbirth. A medical diagnosis made her mother's condition seem more like an illness than a moral fault. Most important, it reassured Carolyn that her mother's decision to leave was a desperate act rather than a deliberate rejection of her children.

Carolyn then considered other factors that might have contributed to her mother's decision to leave the family. Significantly, what she mentioned first was the visible difference between conditions on the farm and in the city. "She wanted better furniture, and she wanted a car, and she wanted to *do* things, I think. She probably wanted more out of life." Maggie did not have to work as hard in the factory as she did on the farm, and households in Endicott had inside plumbing and other conveniences. Carolyn recalled that the Lanes' house lacked basic utilities, just as their farm lacked modern machinery. There was a well right out back, but they worried that it was polluted, so "we'd always bring the water up from the barn." The older girls would carry the water pails in and out of the house because they had a dry sink without a drain. When Maggie was working at E-J during the war, she bought a white ceramic sink and rigged up a drain. But they still had to carry water into the kitchen from the barn.

Sampson had been aware that there were problems in the marriage before his wife finally left. Although Carolyn had no idea what her parents had been arguing about, she recalled being told that he had given her an ultimatum: "I think he told her that if she didn't straighten up, to get out or something. I don't know if that was what really happened, but I've heard him say so afterwards." Soon Carolyn—and everyone else in the neighborhood—became aware that her mother had gone off with a married man. Perhaps her father told the story this way to avoid looking like a weakling or a cuckold. In a later conversation, Carolyn reiterated, "Our mother did take off with another man, although we didn't know that at the time. She'd go down and be gone overnight or something. And then finally this one Sunday she came back and my father said: 'This is it. I told you last night that if you left you weren't going to come back.' So she went down to the road and her brother picked her up. Then it seemed like she just packed her stuff and went. She tried coming back once; she came back for about two weeks and disrupted our whole household that had just got settled down. Then she took off again."

The other man had grown up in the neighborhood and was married to a not-so-distant relative of Sampson Harwood Lane. To his credit, his stepson and ten children were all adults by the time he left their mother, and he waited to marry Maggie until his wife was dead. He, too, had gone to work in the city during the wartime labor shortage; rumor had it that he drove Maggie back and forth to work in his car and loaned her money when she needed it.

What Carolyn remembered from that time, however, was not conflict about her mother's adulterous relationship but rather conflict over finances. Her father handled all the money in the family. When her mother was working off the farm, "I think most of the time she bought things for us kids, because we never had anything until then. I do know that she had enough money saved to pay for having that baby. She wasn't sure whether they [Endicott Johnson] would pay for her" to go to the hospital "because her husband wasn't working at E-J. So she had this money saved. And my father spent it for a horse. One of our horses got crippled and couldn't be used that summer, so my father went and bought another horse. I remember that horse cost $390, and it was about as worthless a horse as there ever was." Normally any extra money the Lanes had was spent on the farm. Her mother accepted that because, Carolyn explained, "that was what their livelihood was; the barn was the first priority." But this incident was different. In later years, Carolyn's mother told her that she had been "very angry" when her husband spent the money she had saved from her earnings to pay for her delivery because "that was her money."

In the wake of her mother's departure, Carolyn had to face down the negative judgments made by others. "Us girls, we lost a lot of respect when our mother took off. But we all just decided" to "never mind what happened, just keep our noses clean and work hard." At that time, situations like her parents' were scandalous. "Because I didn't know anybody that had had a divorce, or their mother had just up and left. We tried to do a lot without making any waves, because we didn't want anybody to think that we were any worse off than we were. It was hard. People talked. They called my mother a whore, and made us all feel bad. You'd hear it; you'd walk into a store" in the village and hear people whispering. Her employer "told me to keep my head up and ignore them. And then I had a gym teacher that helped us out. We'd all our lives been trained to keep up the family name and keep it clean, and that's what we'd always worked for. We all worked together."

In His Sister's Defense

Wesley Saunders, who picked up his sister from the Lane house the day she finally left her husband, saw the dissolution of the Lanes' marriage very differently than Carolyn. He explained to me that Maggie had never complained to her family about how she was treated "until after she left Harwood, and then she commenced to tell about it." The viewpoint he articulated was based on what he had heard from his aggrieved sister as well as what he himself had observed.

Wesley's account had substantial areas of agreement with Carolyn's: Maggie had always loved to work outside, so she did not leave the farm in a protest against overwork, nor did she resent the lack of household conveniences when the family did not have enough money to pay for them. Rather, the conflicts she had with her husband turned on his exercise of authority. Harwood "was a nice, pleasant fellow. I never had no trouble with him. But he and Maggie didn't always agree." Wesley portrayed Harwood as a stubborn man who did everything the hardest way and would not accept others' suggestions. His sister was physically strong and accustomed to having a say in her own life. In his version of the story, their separation and divorce seemed inevitable. Wesley criticized Harwood for "poisoning the children's minds" against their mother after she left him and for severing relationships between members of the two kin groups.

"We've all got our faults, and Harwood had his'n," Wesley began. He thought that Harwood did not do his share of the farm work but left it to his capable wife. "He went to market down there, and he was supposed to be home to milk at 5:30 or something. Well, he didn't get home until 6:00, and Maggie had all the milking to do. So she waited until 6:00 to do her milking, and from then on he was a little later all the while. She got appendicitis or something, and of course Harwood says there wasn't nothing the matter with her. Got Doc Knapp over there and he says she'd got to go to the hospital," so the doctor took her. "When Harwood was in the hospital, Maggie didn't need no help with the chores or anything around there. But when Maggie was in the hospital, Harwood needed some help to do the chores."

"Harwood worked all right; he worked good. But listen, he was kind of a bullheaded-like lad." Wesley told a story about the time the rope on the hay fork broke. Maggie brought it right over to her brother's farm, and he spliced it for her. "I see Harwood in like three or four days, and I says, 'Well, did the rope hold alright?' He says, no, he ain't hooked it up yet. I says, 'Ain't you been doin' any haying?' He says, 'Yeah, but I pitched it off by hand.' Pitching hay over a big beam by hand when you've got a horse fork rope right there! What kind of business is that?" Wesley thought that his sister had to work harder than necessary because of her husband's stubbornness. "She was right out in the hayfield and Harwood was...scratching up hay with his pitchfork." Harwood always did things the hard way, and Maggie "didn't always want to."

Wesley was equally scornful of the Lanes' having to carry their water from the barn. He argued that if Harwood had cleared out the well by the house, which was blocked by tree roots, they could have had water close by. He could also have put a pump in the well in the barn; his grandmother's husband had given him a pump, but he never installed it. "God Almighty," Wesley exploded, "he had a pump hanging right there in the barn, and he was pulling water out of the well with a rope and pail because he didn't have the brains to put the pump in." As Wesley recalled the final rupture of his sister's marriage, she and her husband had argued over the house's water supply. "Maggie left Harwood, finally, and he agreed to have a well drilled if she came back." But she did not stay, so he never did do it. The house finally got running water after the Lanes' son took over the farm. When they got electricity in 1936, Harwood had only a few outlets installed; when I visited the house in the early 1980s, it still had extension cords running every which way.

Wesley readily acknowledged that his sister was unfaithful to her husband. After leaving Harwood, "Maggie went to live with" the other man. "Of course, Harwood wanted to get evidence for a divorce." Under New York State law, the offended party was required to present proof of adultery. "They had wanted to stay" at a hotel "all night, but the manager tipped" them off to the fact that "detectives had come there to catch him and her to bed together. Well, when they come there, he was in there all alone!" Wesley described his sister's dealings with her estranged husband in an ironic tone. "Harwood finally got his evidence on Maggie to get a divorce. [Then] he asked Maggie to come back, see, so she went back and was his wife. That lasted maybe a month or so. He said, 'By god, if you don't behave, I'll divorce you!' She said, 'What're you gonna do for evidence?' He said, by Jesus, he'd got evidence. She said, 'Mr. Lane, your evidence ain't no good.' They were to Endicott, and he sneaks down to Union to see his lawyer.

When he come back she says, 'What did he tell you?' You know, if you get evidence on your wife to get a divorce and then you sleep with her all night, that's no good. The lawyer had told Harwood that." This account suggests that Maggie went back to her husband temporarily for the sole purpose of invalidating his evidence against her. If she were the offending party, she would have lost her right to a share of the farm. Soon "she left him again." When they finally divorced in 1952, Harwood paid her $3,000 for her half of their property.

Wesley vividly recalled others' reactions to his sister's leaving her husband. "The best part of that was, Harwood was down in the A & P store and said that his wife had left him. The manager said, 'She'll be back.' Harwood said, 'No, she won't.' And the A & P manager said, 'Oh, yes, she will; she'll be back.' Harwood said, 'No, she won't.' The last thing when he went out the door, the manager said, 'She'll be back.' It wasn't ten days later that he was a-begging of her to come back. She didn't." Wesley took pride in his sister's ability to remain independent. He taught his sister to stand up for herself. "Well, everybody knew about her" having an affair. "There was a fellow that I used to travel with quite a lot, and he had a girl" away from home. "This fellow liked to throw it into people. So he'd throw it into Maggie" about her being immoral. "So I said, 'Maggie, the next time this guy says something like that, you tell him' so-and-so. She did." Cowed by her allusion to his own illicit affair, he pleaded with her: "'Maggie,' he says, 'since I've got married I've been quite respectable, and, well, you just don't say anything more about this and I won't have anything more to say to you about your past.' He didn't dare open his mouth after that.... It shut him up good." Wesley concluded, "Well, what the hell's the use of seeing your sister having something throwed into her by somebody who's no better?"

In Wesley's opinion, "Maggie was a god-damned good woman to work." She had grown up in a large family of tenant farmers. Their mother, who bore fourteen children, "never made a practice of working outdoors. She done her housework, and she used to have to go to the barn and milk the cows, help milk, when we didn't have a hired man." Wesley thought that his mother "didn't enjoy milking, but you gotta make a living. In order to do that, you have to do a lot of things that you don't like to do." His mother "never hayed it at all," but his sisters did. Wesley recalled one sister driving the oxen to run the hay fork. "Old Will Freeman would come along and was drunk; he'd laugh to see that little damn girl driving them damn oxen." Maggie was the oldest, so she was the first child who was big enough to help their parents in the barn. At that time, "women got right out. Like today, women will get right out and drive a tractor." In Wesley's mind, resentment about being overworked had nothing whatsoever to do with Maggie's decision to leave her husband.

Strikingly, Wesley said nothing about his sister's emotional collapse after her last child's birth. He did, however, contrast the farms and houses in which she had grown up with conditions at the Lanes. The large-scale, specialized farms the Saunders family had rented had more modern machinery, as well as larger dairy herds, than many owner-operated farms. The houses were also better equipped; most had previously been the owners' residences rather than tenant houses.[2] Although the Saunders family never had running water while Maggie was growing up, they did have a pump in the kitchen. They were among the first in the neighborhood to have a radio. It must have been a shock for Maggie to move to the Lanes; the farm was smaller in scale, and the house was ill-equipped. Wesley thought that his sister worried about not having basic necessities. Maggie had told him that once she "wanted Harwood to give her some money to go buy groceries with, and he wouldn't do it," even though it was only a few dollars. She borrowed this sum from a neighbor who had overheard their exchange, which was quite unusual. In his account, it was not clear whether Harwood had the money his wife needed and refused her request or simply had no money.

In a letter written to me after our interview, Wesley acknowledged, "Well as you know we all have our faults and it is not very nice to rake up all a mans faults and his fathers familys faults." In defending his sister from her husband's accusations, however, Wesley had found it necessary to criticize him. This acrimonious separation and divorce continued to divide the Saunderses and the Lanes almost forty years later. Still, he was glad that the Lanes had enabled me to hear his point of view: "It would seem to me that if the Lane family Harwood children did not want that this was told one of his children would not of ever brought you over to see me as some of them had a good idea of what I would tell you or write."

An Older Daughter's Perspective

Clarice, Maggie's oldest daughter, disputed many of the criticisms that Wesley made of her father. She contended, for example, that Sampson Harwood did not linger in town after taking the milk down; he would do his egg route, buy groceries, and fill up the truck with gas before coming home in the afternoon. Once a week, he made a payment on the mortgage and paid his life insurance. When he was late, her mother "might start the chores, but there were hard milkers and some that kicked that only" her father milked. Her parents both milked, but they "never milked at the same time, even if they were home all day." "When the work load got too much for them, they had day help....There were several young men and boys that their parents could spare for a day that helped when Maggie wasn't able to." Clarice shared Carolyn's and Wesley's perception that Maggie "loved to drive horses and work outdoors more than in the house." In her memory, her mother "seldom plowed, dragged, or used the reaper, but could if the need came. She did mow the hay while he traveled around the fences and [cut] any spears left standing. They worked well together and each one did what the other didn't. We never heard any complaint."

Clarice offered no explanation for the breakdown of her parents' marriage. Instead, she recalled their cooperative working relationship. They divided the labor in what seemed to her an equitable manner, although it did not conform to the dominant culture's division between male and female domains. In the morning, Harwood "would rise first, start the fire and put the coffee over, go to the barn and start the milking, then call Maggie to help. She would call one of the older children to sleep down stairs so as to hear the baby if it cryed." After the milking was finished,

> he would repair machinery, feed chickens, pigs, sheep and other animals while she took the cows to pasture. Then she would wash up the milk pails and strainer, and get breakfast and call him from wherever he might be. She would get the dinner started while he was hitching the horses up. [With] wood stoves, you could do this by putting the meat nearest the fire, vegetables farther back. On days it rained she would clean house, wash clothes, and bake—or on days there wasn't a job to be done. This she did until she went to work in E-J. Then sometimes she didn't go to the barn. The girls were big enough to take her place then. Once when we had a load of hay yet to put in, and it was going to rain before morning, she came home (probably tired too), changed into her farm clothes, and helped us get it in. What one more hand can do sure helped. She was strong and kept herself that way.

Her mother demanded the respect she deserved. "There may have been some words between her and the girls," Clarice admitted, "as she wanted them to do what she said right then. She expected them to do the work like she could."

While acknowledging the disrepair and lack of conveniences in the Lane household, Clarice ascribed them to practical problems rather than her father's unconcern. She admitted that her father "was set in his ways...and only changed some in later years." But, she maintained, there was logic in his decisions about what to do and what not to do. Other "people had to carry water further than we did." In her view, the family's poverty, rather than her father's laziness or neglect, was responsible for their difficulties. The Lanes' hold on the land remained insecure, both legally and financially. Her parents "did not know for sure if" his grandmother's husband "was going to sell them the farm or not, so they bought the one above them....They rented [out] the house, stored hay and kept heifers in the barn, and used the land. It joined the one they were on. They sold it somewheres near 1934....At one time they were paying on two places. Then the Depression!" The family's lack of money did not mean that they were deprived. "They could raise enough to eat, just needed cash for seed, gasoline for truck, clothing and doctor. The cash came from milk and eggs." Clarice remembered "them telling of how" she, at four years, and her sister, at two and a half, "went to the chicken coop and broke an egg each that day. This is one of the few times he [their father] did any spanking. Maggie was strict—she would spank or switch the older ones for most any thing. She never tried to talk to us (if she did, she may have found that it didn't work well in our cases). She was hard, but when we needed Love she had that too."

After their parents' divorce, Clarice said, "the older girls would get their boyfriends to take them to visit their Grandmother Saunders. The children loved the grandparents same as always, but felt true to their father. He never asked us not to go, but when we told him of the visits he wasn't overly thrilled." Clarice had the impression that their mother's second husband "didn't want her to write or see us. (Maybe he was afraid she might get homesick and want to leave.)" After he died, Maggie moved back to the area and got to know her youngest child, who had no memories of her. Then she remarried and moved away again. When she came to visit, "there was much happiness...for all of us," Clarice concluded.

Clarice's account barely acknowledges the pain that her mother's desertion must have caused her. She married just a year later, at the age of seventeen, but was divorced soon after her child was born. One sister "had to get married" at the age of seventeen. Another sister married at nineteen, had two children, and was divorced. Only Carolyn's marriage endured. Nonetheless, the family history Clarice wrote is resolutely upbeat. Of her father, she said: "He was easy going. Laughed a lot. The happiest day in his life was when he had a Son." Of her youngest sister, Clarice wrote: she "grew up with the mothering of her sisters and brother. Her father being both parents did his best."[3]

Maggie's Departure as a Sign of Change

The story of the revolt of Maggie Saunders Lane, as it was variously told by her daughters and her brother, attests to the fact that even a woman who loved working in the barn and the fields and preferred being outdoors to doing housework might reject the position in which she found herself. After twenty-two years of marriage, Maggie despaired of ever improving her lot and enjoying life on the farm that she and her husband had built together. She might have stayed on indefinitely, fatalistically regarding her situation as unalterable, had she not taken a paid job off the farm during World War II. In the factory, she made money she felt she had a right to regard as her own. Maggie spent some of her earnings on a sink with a drain and tried to persuade her husband to have a new well drilled so they could have water in the house instead of carrying it from the barn. When she unexpectedly became pregnant, she planned to use her savings to have her baby in the hospital, not at home like her first five. At that point, however, her husband took her money and spent it on a broken-down horse. What was at stake here was not the relative priority of what the husband thought the farm needed and what the wife wanted for the house, as the issue was formulated by Mary Wilkins Freeman. Maggie, like most farm women, knew that it made economic sense to put the barn ahead of the house. The crux of the matter, rather, was who controlled the money she earned and who made the most important decisions about her life. After giving birth, Maggie sank into a deep depression. She left her family in order to save her sanity. As adults, neither of her daughters blamed Maggie for this course of action, even though at the time they suffered from her absence and were shamed because of her alleged immorality. When her husband's authority over her became unbearable, Maggie Saunders Lane stood up for herself.

That she was able to break free from a marriage she found oppressive reveals how much had changed by World War II, when rural residents went to work in the city every day. That her rebellion cost her so much, especially her relationship with her children, is a measure of her desperation as she struggled with depression. Despite her older daughter's efforts to convince herself that in the end all was well, this story has no resolution. Instead of the scene that Mary Wilkins Freeman depicted, with an older couple, their married daughter, and her husband living happily ever after in a spacious but snug converted barn while the cows are sheltered in their old house, this story leaves us with a bitter man living alone in a dilapidated house on a run-down farm and doing everything the hardest way. The energetic, resourceful woman who knew better was gone. Maggie was always ready to milk cows and rake hay; she did housework only when there was nothing else to do. But she insisted on having a voice and improving the quality of life for herself and her children. When her husband refused to listen to her wishes and interfered with her getting proper care in childbirth, violating her personhood in a particularly offensive way, she left. The revolt of Maggie Saunders Lane, then, was not a mother's reaction to her husband's putting the barn before the house, to being overworked, or even to a pervasive lack of spousal communication, but a woman's assertion of the power to shape her own life, within her marriage if she could, but independently if she must.

Other women with experience of both farm work and off-farm employment bargained with their husbands so they could do the work they preferred. Both immigrant newcomers and locally born women described their explicit negotiations with their husbands. Although they made compromises because of difficult economic circumstances, they were able to lead lives they found more fulfilling than frustrating.

"I'll Work in the Factory, and You'll Take Care of Your Farm"

Joanna Chesky Misa and her husband, Stanislaw, moved to a farm in the town of Maine in 1926. She chose to keep her job in the city while her husband worked full-time on the farm.

Joanna Chesky was born to Polish immigrants in 1899 in New Bed-
ford, Massachusetts, where her father worked in the textile mills and
her mother ran their small farm. After her father died when Joanna was
young, her mother raised six children alone. "She had four acres, and she
kept three or four cows milking. She sold her milk, and then she'd help out
with the different neighbors if they needed help, if they needed corn cut
and like that. Things were done more or less by hand, not by machinery
like now, so...women, if they wanted to help out, they could make a dollar or two." Her mother also "had chickens, and she'd take her eggs and
a chicken" and sell them in town. Joanna recalled eating rye bread with
bacon drippings, which was delicious as well as filling. She went to school
until she was fourteen, as the law required, but for the last year she lived
with and worked for an older woman for $2 a week. Her mother refused
to allow her to go to high school, even though her employer would have
kept her on: "She thought about it, that she's [Joanna's] going to get married, and what is that high school going to give her?" So she got a job as
a spinner in the mill.

Joanna married in 1917, when she was not quite eighteen; her husband
was almost twenty-five. Stanislaw Misa came from Galicia, a part of Poland
then ruled by Austria. They had met at a Polish church picnic when she
was fourteen, but had to wait to marry. He worked as a weaver in the mill,
and she had no idea that he wanted to be a farmer. "I'd have never married him," she exclaimed; "I would have run more than I ran. I couldn't see
working in the dirt! No, my plan wasn't on the farm. I was gonna work in
the mill, be a city girl." But her husband decided to move nine miles outside
of town to "a stinky little farm." Joanna thought that farming was too hard
and they would always be poor. She "didn't argue with his decision," however, because the money he spent was his own. "No, I didn't buy it. I had no
money invested in it, and I didn't feel that it belonged to me." At the time,
"I felt, well, he's older than I am; I shouldn't argue with him. If he wants to
go on the farm, well, we'll have to go on the farm. I got pregnant, so I had
to quit my job in the factory and go on the farm with him." After the end of
the war, the cotton mills were going out of business and the market for farm
products collapsed, so "we lost everything."

It was also her husband's decision to move to New York state. Stanislaw
joined his brother at Endicott Johnson.

> But he didn't like to work in the factory, my husband. He worked in the silk
> mill [in Binghamton], but he didn't want to stay because the daytime pay was
> too small. So he decided he wanted to be on the farm. I had a job in town,
> and he worked at night in the silk mill until we got up enough money to buy
> a little farm. He thought he could do good on the farm. It didn't make too
> much difference if I liked it or not. I didn't care for it too much, his plans.
> But I thought maybe it would work. We had two children, and they would

have a place to run and play and not get into trouble. And he'll be home with them all the time, and I can work.... So that's the way it ended up; we bought the farm.

As soon as he "could scrape up a little money so he could do a little more on the farm," he quit his job. Joanna recalled how they decided to allocate their labor. "He liked farming, and I didn't like it. So I said, 'I'll work in the factory, and you'll take care of your farm.'" "My husband was home on the farm, and he took care of the kids and done his farm work. So that's how we worked it. Maybe it was unusual, but it worked out better for us."

The Misas bought a 61-acre farm on a hill above Union Center. "It was big enough for him to take care of; in fact, maybe it was a little bit too much for him to take care of alone." Their four daughters "did whatever they were able to do: hay, work in the field, raise cauliflower and cabbage. That was his specialty, cauliflower." Their daughter Dellina, who was present during our interview, added, "Dad plowed when we had horses, but after we got the tractor I plowed, because I loved that." They also raised heifers for nearby dairy farmers. "We never had a dairy, although some of them would freshen and Mother would make cheese and butter" for their own use. During World War II, they sold some dairy products. The Misas kept chickens and sold eggs as well. Joanna brought their farm produce to work with her and sold it to her co-workers.

It took Joanna several years before she found what she regarded as a satisfactory job. "I was determined to get into E-J, no matter how, because I could get benefits for my family. I thought, well, if I don't make high wages, so what? I'll get the benefits, and I'll be better off." At Endicott Johnson, she eventually got a position with acceptable conditions and reasonable wages. She stayed there eighteen years. When I asked her who made more money, she replied that her husband "didn't worry who made more money; he wanted his farm, and that's where he was gonna stay." He had no fear that she would assert herself because she was the family's wage-earner. "He knew I would never go out to be boss; I wasn't that type of person to be bossy or aggressive or arguing. I never argued with him because I always lost, I never won out. So I let him do just what he wants to do; how it came out, I just went along with it. As long as we got ahead, not behind again. He was all in favor of my working on the job and bringing in the pay so that I can pay the taxes. Taxes were paid, all the insurance taken care of, and the payment on the place; everything was taken care of" by her wages. "How much he made, well, that's how much he made." It never occurred to her that because she was earning at least half of the money she ought to have an equal say in the decisions. "We just pooled our money together, and whatever bill came we paid, and whatever was left, well," they made do.

When they moved to the farm "the old barn wasn't too good. He wanted a good barn, so he could raise calves and have shelter for them. So whatever money he had left after the bills were paid, he'd figure to put toward building the barn." He cut lumber on their land, bought cement, and "built a barn, a big, beautiful barn." Some of his friends from the city helped him nights and weekends. The barn definitely came before the house: "We didn't plan on the house then."

Stanislaw built their house much later, after all the children were grown. Joanna found this order unremarkable, even though the farmhouse was old. "We didn't have no running water. In fact, we didn't have no water, period. I had to carry the water into the house and out. I had a sink, but…there was no drain." The well was down a slope about one hundred feet from the house. "It wasn't exactly a hill, but it was steep enough so that in wintertime you'd come up with half a pail of water." They had no electricity until 1937. Without even a hand-operated washing machine, they did the laundry on a scrub board. The girls helped with the housework as well as on the farm. Dellina recalled cutting wood for the stove: "As soon as we got old enough to handle that cross-cut saw, he'd say if you want to stay warm, come out and work." Her father did not take on the household tasks while her mother was at work. "We lived close to the school, and I recall running home at noon. He wouldn't have lunch ready; we had to fix our own lunch. But he was there always; no matter when you came home, he was there."

In retrospect, Joanna thought that she deferred to her husband's opinion too much and they might have had an easier life if she had spoken up more. When they married, "he was older than I was, and he was a man of the world, he had traveled, and I…went nowhere until he come along and dragged me." Although she spoke English better than he did and was more familiar with American customs, "I didn't show my authority in any way, and so he just took over." "I felt that he was more right than I was, that he had more knowledge about everything than I did. He just led the conversation, and he knew that more or less I would go along with it. He would express himself in a way so I felt that he was right and he knew what he was doing, what he was talking about, and if we'd go along in that" they might succeed. She was always convinced that differences of opinion between spouses were not useful: "If I'd go one way and he'd go another, we're not gonna get anywhere. But if I follow him, maybe we'd come out somewhere. And it always did. It was hard. Maybe if I'd had something to say, or we could have talked it over, maybe we'd have found an easier way to get to the same point and not have had it so hard. He didn't always have the best way. He'd try; he thought it'd be the best way; but sometimes it would end up that we was doing it in the very hardest way. We'd get there, but it would be the very hardest way. I think that maybe if I'd pushed to talk over with him, maybe it wouldn't be such a hard life. But I didn't push to talk things over, and I just went along.…Oh, I didn't want to argue; I'd rather give in than argue. I never won! What's the use of arguing?" If she could live her life over, Joanna "would show him that I've got as much, in fact I've got more experience in this country than him, and I know how to deal with it better than him."

Joanna deferred to her husband on the farm but determinedly held on to her job in the city, which she much preferred to farm work. The division of labor she and her husband worked out satisfied both and supported the family. Over time, however, she became convinced that a more equal partnership might work better. The example of her daughter's married life was instructive. Dellina, like many American-born daughters of immigrants and their peers from families with deeper roots in the Nanticoke Valley, was more assertive than her mother had been. Another woman of Dellina's generation and class position told that sort of story.

"We've Always Talked It Over and Made Decisions Together"

Lydia Bradley Meyer, who was born into a native-born family in 1925, grew up on Tiona Road. The Bradley family exemplifies the combination of farming and wage-earning that both native-born and immigrant families adopted between the wars. Although Lydia was not interested in farming, she and her husband eventually bought a farm of their own. The labor arrangements she and her husband worked out varied over time; they took turns having an off-farm job and managing the farm operation. Although Lydia preferred working for wages to dealing with the dairy cattle, what was most important to her was the cooperative relationship she had with her husband. Whether she contributed her labor or her earnings, she felt rewarded by their family's ability to make a go of it in a way that suited them.

In her childhood, Lydia's family moved back and forth between the country and the city. After the Bradleys bought a small farm in 1930, her father kept his full-time job at Endicott Johnson. In 1937–1938, they returned to Endicott, and both parents worked in the factory. Then they traded their house for a larger farm on Tiona Road and kept their jobs. "They did the farm work when they weren't working in the factory." In Lydia's experience, "most women helped, in the barn and even in the fields, when they could."

When I asked Lydia why her parents decided to farm even though they never had enough capital to farm on a large enough scale to enable either of them quit working in the factory, she replied that "Dad was brought up on a farm in East Berkshire, and I think he kind of wanted to get back onto a farm." "My mother had come up from near Stroudsburg, Pennsylvania," to work at E-J. She had grown up "out in the country, but my grandfather didn't farm it too much; he had just a little place. It was definitely out in the sticks!" Lydia's mother learned to milk after they moved to the farm in 1930. Her father's mother had always gone to the barn and helped out with milking and chores: "She had eleven or thirteen children, but I'm sure she did her share." So Lydia's father was accustomed to the women working right along with the men.

On their first farm, the Bradleys milked between ten and a dozen cows. "My mother worked in the barn and churned butter" after separating the milk with a hand-operated machine that was kept right in the kitchen. "We did not have running water at that time; we had a pump in the kitchen." So butter-making was a laborious process. "My Dad would take some into the factory and different friends and relatives would buy the butter." Lydia assumed that the proceeds were her mother's, although "I don't remember how much she got for it." Her mother also "helped my Dad" in the fields. When I asked what "helping" on the farm involved, Lydia spoke for herself and her sister as well as their mother: "Well, we drove tractor and truck in the hayfields, mowed hay, plowed and dragged and so on and so forth."

On the larger farm, they milked between twenty and thirty cows. "My Mother and Dad usually, when I was in high school, did the milking, but it was my responsibility to wash the milking machines and the milk pails before I went to school. Then my sister did the same thing after I left home" to get married in 1943. From 1938 on, they had "a regular hired hand"; labor was cheap during the Depression. The Bradleys concentrated on dairying and sold fluid milk. They produced their own hay and cut corn for silage. For a short time during World War II, her father left his job "and just worked on the farm," but then "he bought a milk truck" of his own "and had a milk route; he took the milk cans and delivered them to Owego."

Figure 9. Pitcher twins taking milk to H. A. Niles Creamery in Maine village, ca. 1902. Photographer unknown. *(Courtesy of the Nanticoke Valley Historical Society.)*

The Bradleys produced less of their subsistence than families who farmed full-time and had no income from wages. They had a vegetable garden and Lydia's mother canned, but the garden was "not a big one" and they sometimes bought canned goods to supplement what it supplied. They also kept chickens and grew potatoes for their own use. Their house had no electricity and few conveniences. First her mother had "a washing machine that turned with a hand lever, and then she had a gasoline engine on the washing machine."

Lydia had the impression that the women in their open-country neighborhood all went to the barn with their husbands. One of her sisters-in-law did not go to the barn, but she had been raised in town. The other women "all worked in the fields unless there was a crew of men there, too." Lydia thought that they did more of the field work when their children were young, and less when their sons were old enough to take over. Once the Bradleys lived on the larger farm, they got to know the immigrant families who also lived on Tiona Road. They got along especially well with the Polish family who farmed next door. The men all changed work during haying, corn-cutting, silo-filling, and thrashing, and the women helped her mother cook dinner for the men. It did not seem to her that immigrant women were any more likely to work outdoors than native-born women were. "If you were a farm family, the women worked. I think maybe the immigrant women had more stamina, or maybe they had more determination. But I think they all worked."

While Lydia's narrative focused on her parents, her viewpoint was strongly colored by her own experience. She did not choose to become a farmer's wife: "No, I swore I'd never marry a farmer!" She laughed ruefully. "I'd had enough of it." But when I interviewed her she had lived on a farm for thirty-three years. She recounted how she and her husband, Henry, came to buy a farm at the end of World War II. "Well, we felt it was time to get something of our own. He wanted to farm. His father was farming, too, and he liked farming and wanted to be his own boss. So that was it." Lydia and Henry were familiar with the place they bought, as both had lived nearby with their parents. "We knew it was small," but "there was land available for renting if we needed it." Eventually they bought his parents' farm as well.

For the first couple of years, Henry kept his job in town and Lydia was the day-to-day farm manager. They began with a dozen cows and slowly built up their herd. "I did the milking in the morning and took care of things, all that I possibly could." He milked in the late afternoon. "We worked until dark a good many nights, both of us." When I asked Lydia what made her willing to be a farmer, she said laughingly, "I don't know, crazy, I guess." She thought "it was good for a family to be brought up on a farm." Their three boys enjoyed being outdoors and were expected to help with chores. "They all worked on the farm as soon as they were old enough, sometimes a little bit too young, maybe. But they worked through high school. None of them wanted to be a farmer, either, like I didn't!" Lydia laughed again. Her ambivalence about farming was audible despite her acceptance of her husband's wishes.

Lydia described her work both on and off the farm in terms of its contribution to the enterprise. "We had been here five years, and Henry wanted to build a 100' by 60' addition onto the barn, to put more cows in. At that time I was getting $100 a month" from the milk check "to run the house on. And so I said, okay, if you want to build, I'll go to work at E-J, and you can use that $100 to pay for the barn. And that's what I did; I worked four and a half years." Her wages "usually went for groceries and household expenses," while all the money from the milk checks "went into the barn, for the farming equipment and the operation. Once in a while I didn't have quite enough for everything, but most always what I made was enough for household expenses." She was integrally involved in decisions about investing in the farm. "We always talked it over first, and decided what we were gonna do." Lydia felt like a full partner in the business: "We've always talked it over and made decisions together."

She quit working off the farm when her third son was born. "Then I stayed home, took care of him, and did what I could on the farm. In the summer, I'd still go out and help all I could, even with the baby; he'd ride with me in the truck, and I'd drive bales of hay down or whatever. Then when he went to school I started working part-time in the school cafeteria for three or four years, just a little added income. I'd still be here mornings and evenings to help Henry. Then I went to work driving school bus, and did that for fourteen years." Lydia emphasized the fact that when she went back to work full-time, Henry was working on the farm full-time and was home to supervise the children. When I asked why she, rather than her husband, worked off the farm to bring in extra income, she declared, "Because I liked it better, let's put it that way," and laughed. It seemed to me that she wanted to avoid saying anything too negative about farming. "I guess I just plain got tired of being around the cows all the time."

When they moved to their own farm, the house had electricity, running water, and a bathroom. This was an improvement over where she had grown up; "they didn't get electricity on Tiona Road until 1938." In other respects, however, the house was not so different. It lacked central heat, kitchen cupboards, and a downstairs bathroom. Enlarging the barn definitely took priority over improving the house. "It was tough getting along. Everything went on the land and into the barn for a good many years. There were times I resented it, but not too much. Because I was working there also" and knew what was needed. "Besides, I've been the bookkeeper since we moved here. So you kind of went along with it. You know what has to be done."

When I asked Lydia to compare her life with her mother's life, she responded, "I don't know as it's been any easier, actually. The manual labor has maybe been a little bit easier, working out in the fields. But you're still tied down." When they go away for important family events, "you worry about things. But we have been away, these last few years, where my mother never was. Other than that, I don't think there's that much difference."

The most important difference between herself and her mother, which Lydia did not emphasize when I asked her to compare their lives, became clear when we talked about who made decisions in farm families. In the Bradley family, Lydia said, "my father was boss, and mother went along with him. They did talk things over, but I think he made the decision even after they talked it over, and she'd go along with him. He never said too much about what went on in the house; she was sort of the boss in the house." As far as the children were concerned, "Father was the disciplinarian. Mother would get after us every once in a while, but I don't remember my mother ever spanking me, but my father sure did!" Lydia had the impression that relationships in other families were similar. "I think in that era, most families were male-dominated, more so than now. I think that was just the way it was and you accepted it, that was it."

Lydia said that when she was young she decided that she wanted a husband who would treat her as more of an equal. "I made up my mind that I wasn't going to sit down and let someone yap at me without yapping back!" She laughed. "So probably a lot of the other [girls] had the same idea. But that was the way of things among the grown-ups when I was a girl." "My father drank quite a bit. I never knew it till I was fourteen. And when he did drink, my mother never argued with him; she would just sit and take whatever he had to say. I swore I would never do that. And I got too much the other way; I'm much too mouthy!" She laughed and then explained that even as a young woman she stood up for herself and for her mother, although it was scary. "The only time he would ever say or do anything abusive at all was when he was drinking," Lydia emphasized. "He was a great guy, really, awful good to us all. But we had to mind. And when you didn't, you got it!" Lydia tried to avoid conflicts with her father when she was growing up. "I'd speak up once in a while, if he had been drinking, and not very often, I would speak back." But if he forbade her to do something, "I wouldn't do it. And I wouldn't argue, either. There were very few times, even when he was drinking, actually, that I really dared to speak up too much." Lydia vowed that she would never marry a man to whom she could not freely speak her mind.

In marrying Henry Meyer, Lydia chose an easy-going man who could be trusted to listen to her and value her opinions, as well as enjoy her company. Over the course of their married life, she and Henry negotiated the most crucial decisions about how they would make a living. Lydia recognized that she and her husband had differing interests; he loved farming, but she found it confining and was bored by milking cows and doing chores. Yet, when it made economic sense, she acted as the full-time farm manager while her husband held an off-farm job. In her mind, the major advantage of farming was that it enabled her to combine income-producing work with childcare. When their financial and family circumstances allowed, however, Lydia preferred working for wages and supporting the household while the milk checks were reinvested in the enterprise. In either capacity, she was a full partner. What was most important to her was not whether or not they farmed, or even whether she worked on or off the farm, but her partnership with her husband, which was based on mutual respect, the recognition of their differing personal preferences, and a deep commitment to negotiating agreements that satisfied both.

PART IV

ORGANIZING THE RURAL COMMUNITY

9

![decorative ornament]

Forming Cooperatives and Taking
Collective Action

At the turn of the twentieth century, Nanticoke Valley farmers had
a vital tradition of forming cooperative associations, membership
organizations through which they not only purchased fertilizer and
feed but also processed and marketed their products. Bargaining collectively
and excluding the middlemen who would otherwise profit from their trade
made farming more sustainable economically. It also ensured that farmers had practice in conducting organizations and handling financial affairs
democratically. Granges—local chapters of the Patrons of Husbandry—had
existed in this community since the 1870s, and in the early twentieth century, cooperative creameries in Glen Aubrey, Maine, East Maine, and Union
Center processed milk from members' farms.[1] With the shift to fluid milk
production for the New York City market, however, farmers were at a dis-
advantage because large milk companies controlled the prices paid to farmers as well as by consumers. In the mid-1910s, Nanticoke Valley farmers
joined the Dairymen's League, which united producers in the New York City
milk shed. First, the Dairymen's League negotiated better prices with milk
shippers and processors; then, it established its own processing facilities.[2]

The history of the Broome County Farm and Home Bureau, as well as the
local branches of the Dairymen's League and the Grange League Federation,
demonstrates how rural men and women organized themselves to solve the
economic and social problems they faced. Farm people drew selectively
on the resources available to them through the New York State College
of Agriculture at Cornell University,[3] but they defined their own goals for
their family farms and communities. Nanticoke Valley farmers followed

the advice of experts as they limed their pastures, improved their dairy herds, and culled their poultry flocks. At the same time, they undertook collective action to redress the imbalance of economic power between themselves and the capitalist firms that purchased, processed, and sold their products and from which they purchased inputs.[4] Working through the Grange and the Farm Bureau, Broome County farmers established both producers' and consumers' cooperatives.

County- and state-level organization was crucial in enabling farmers to control the marketing of their products. At the local level, however, farm people adapted these large-scale organizations to their customary modes of social action. The Broome County Farm and Home Bureau was organized in small groups centered in particular localities. In the Nanticoke Valley, Glen Aubrey, Maine, and Union Center all had active Farm and Home Bureau units. Rather than confining their concerns to agriculture, these groups took part in a wide range of local affairs. They cooperated closely with other organizations, especially the churches, and organized community-wide social and recreational programs.

Most important, the Farm and Home Bureau had a coordinate structure that allied men and women in the improvement of family farms and rural life. In contrast to the experts in farm management and home economics who advised the Farm and Home Bureau, who operated within an ideological and organizational framework that was gender segregated, conceiving of the farm business as separate from the family home and defining a clear

Figure 10. Women, men, and children taking a break from threshing, Maine, New York. Undated photograph by Lee Loomis. *(Courtesy of the Nanticoke Valley Historical Society.)*

boundary between the proper domains of men and women, the Broome County Farm and Home Bureau ensured the active participation of women and promoted cooperation between women and men. At the local level, Farm and Home Bureau groups were closely related. Husbands and wives worked together as farmers and community activists. Home Bureau groups sponsored events for men as well as women and children, and Farm and Home Bureau units undertook joint projects to improve their communities. Farm people, in sharp contrast to most professionals in the agricultural sciences, agricultural and home economics, and rural sociology, integrated the activities and concerns of women and men.[5]

Founding the Broome County Farm Bureau

The Broome County Farm Bureau, established in 1911, was the first community-initiated and locally supported agricultural extension agency in the United States.[6] The process of its formation illuminates both the commonalities and the divergences in the perspectives and interests of farmers, experts, and businessmen during the second decade of the twentieth century. The report of the County Life Commission, chaired by Liberty Hyde Bailey, dean of the New York State College of Agriculture in Ithaca, alarmed Byers H. Gitchell, secretary of the Binghamton Chamber of Commerce, who recognized "the interdependency of city prosperity and country prosperity." Binghamton was the leading agricultural service center in Broome County; its mercantile and manufacturing enterprises purchased and processed farm products; produced and purveyed farm vehicles, tools, and machinery; and furnished the countryside with consumer goods. By the turn of the twentieth century, new industrial plants were being built in the villages founded by the Endicott Johnson Shoe Company, rather than in the older municipality. Gitchell sought to stimulate the city's growth by strengthening its economic ties with the hinterland. In summer 1910, he proposed "extending to farmers the same opportunities for cooperation now enjoyed by the business men of this city."[7]

Enlisting "three successful farmers from the adjacent community, one wholesale grocer, and a certified milk-producer," Gitchell established a "committee to investigate farming conditions and to report on what action should be taken to develop farming opportunities."[8] Aided by experts from the New York State College of Agriculture and the United States Department of Agriculture (USDA), the committee toured Broome and the adjacent counties and concluded that "something should be done to inform all the farmers of the opportunities afforded by agricultural science and good farm practice."[9] According to this account, the difficulties that farmers were experiencing were not the result of the poor quality of the land; their problems could be solved by the application of scientific methods and business management. Many farmers agreed "that study, well-directed labor, capital, and confidence would restore productivity to low-priced farm lands. The committee concluded that it would be necessary to analyze the local agricultural conditions and then demonstrate local opportunities for improvement by concrete examples of success." The statement that "farm management rather than population was seen to be the main question involved" underlines the narrow framework of this inquiry.[10] The committee, like the USDA and the federal Census Bureau, accepted the continuing decline in the number of farms as an inevitable result of the consolidation that accompanied agricultural progress.

At the same time that Gitchell's committee was doing its work, George A. Cullen, traffic manager of the Delaware, Lackawanna, and Western Railroad (DL&W), was seeking a way "to help farmers along its lines."[11] The DL&W extended from Broome County south to New York City, west to Buffalo, and north to Syracuse and Utica. Along with the Erie Railroad, it was a major carrier of farm products and manufactured goods to and from the region. Like the Binghamton Chamber of Commerce, it had a direct interest in the prosperity of agriculture in the county. Cullen explored the possibility of establishing a demonstration farm or experiment station that would teach farmers scientific methods. Consulting with experts in Ithaca and Washington, D.C., Cullen was advised by William J. Spillman, head of the Office of Farm Management at the USDA,[12] that it would be more effective to employ a county agent to work directly with farmers. The USDA already had agents in the South, so Spillman was speaking from experience when he counseled that educational work among farmers ought to be "local, concentrated, and continuous."[13]

The railroad, the Chamber of Commerce, the USDA, and the New York State College of Agriculture agreed to establish the Broome County Farm Bureau in 1911. The College of Agriculture and the USDA provided educational assistance; the USDA, the railroad, and the business organization financed the project. The objectives of the Farm Bureau were

> To undertake propaganda work in the agricultural district in the vicinity of Binghamton, New York, to make an agricultural survey of the territory, to study the farmer's problems, to find their solution by a study of the practice of successful farmers, to study the relation of types of farming to local conditions of soil, climate, and markets, to demonstrate the systems of management used by the successful farmers of the district, to conduct demonstrations with farmers, to carry on educational work through the media of institutes, to advise with farmers individually or otherwise as to the best methods, crops, cropping systems, stock, labor, tools, and other equipment.[14]

This statement of purpose makes clear that the Farm Bureau was intended to be primarily educational. Once the agent had surveyed local conditions, he was to teach farmers scientific methods of agriculture and efficient management. It was necessary to involve farmers in the work because they would be convinced of the benefits of new practices only by seeing them successfully employed by people like themselves. But farmers were not to be involved in the selection of the problems for study or the definition of criteria for acceptable solutions. Agricultural economists assumed that the structure of the market was a fixed condition, similar to the soil and climate, to which farmers had to adapt, rather than challenge or change.

John H. Barron, the first Farm Bureau agent, began his work in March 1911. He toured the county, inspecting agricultural conditions and making contact with farmers. Those who expressed interest in learning new methods were asked to organize meetings and sponsor demonstrations in their localities. These were often followed by the formation of Farmers' Clubs, which met fortnightly in district schoolhouses to study Cornell bulletins and exchange ideas about common problems. Barron reached out to established farm organizations, especially the Grange, and used customary modes of rural social organization to involve farmers with the bureau.

The active participation of farmers began to shift the priorities of the Farm Bureau. In addition to "serving as a clearing house for the agricultural progress of the region," teaching farmers to prune their orchards, cull their poultry flocks, and lime their pastures, it became a vehicle for direct action. As the first history of the Farm Bureau delicately put it, "As a result of a successful campaign for lower freight rates on lime, the use of that commodity for ameliorating soil conditions was stimulated."[15] This careful phrasing cannot conceal the significance of this development, although it attempts to depoliticize it. The Farm Bureau agent's educational effort to convince farmers to lime their land immediately encountered an economic obstacle; lime cost more than many farmers could afford because the bulky commodity was expensive to ship. The Farm Bureau responded by campaigning for lower freight rates, which it ultimately obtained from the DL&W. Although the railroad might have made some friends by its willingness to serve the farmers along its lines, farmers went beyond petitioning for concessions from a business benefactor. Following the example set by the Farmers' Alliance and Populist movement in the late nineteenth century, Broome County farmers had served notice that economic issues must be addressed.[16]

Barron managed to involve some "practical farmers" with the Farm Bureau and demonstrate that it could serve their needs. Nevertheless, many farmers remained dubious about the new agency. According to the later history of the Farm Bureau, "Though the program was ambitious, farmers were more or less indifferent. They felt that something was being done for them—some thought to them—in which they had little or no part. They believed that business interests were acting from selfish motives in getting the farmers to produce more food while the farmers were interested in getting more money for what they produced."[17] Broome County farmers' resistance amounted to more than passive indifference or the proverbial skepticism of practical farmers toward "book farming." They believed that the Farm Bureau was undemocratic, controlled by business interests and experts rather than by farmers themselves. They thought that their own economic interests were in conflict with those of the capitalist firms that controlled the terms of trade in farm supplies and agricultural products; these middlemen profited at farmers' expense. They feared that the agricultural experts' program for increasing productivity would serve business rather than farmers because the prices of farm commodities would continue to decline as output rose as long as farmers had no control over the market.

This critique was made not only by Broome County farmers who remained aloof from the Farm Bureau but also by those with whom Barron came into direct contact. Evidently Barron was able to appeal to farmers who were critical of business interests as well as to those who felt comfortable with members of the Binghamton Chamber of Commerce. As the later history of the organization puts it, "County Agent Barron was a member of the Grange, and this, together with his practical interest, saved the day. Had he not been a farm-reared boy with a practical turn of mind, the cause most certainly would have failed."[18] Farmers accepted Barron because he shared their background and perspectives. Through his participation in the Grange, he advocated that farmers organize economically and politically to remedy the problems they faced.

Organizing for Economic Cooperation

Its close relationship with the Grange proved crucial to the development of the Farm Bureau. In 1913, a Farmers' Convention called by the Farm Bureau formed the Farm Improvement Association of Broome County (FIA), an independent organization controlled by the farmers themselves. The proposal to establish the new group was supported both by Lloyd Tenney, state leader of the Farm Bureaus at the College of Agriculture, and by James Quinn, master of the Broome County Pomona Grange. According to E. R. Minns, "the conviction grew in the minds of the State Leader and the local agent that in order to make the work of this bureau most efficient and successful, there must be behind it the organized sentiment of the farmers for whose assistance it was formed."[19] His 1914 account omits the farmers' critique of the Farm Bureau that was included in the post–World War II history, but indicates that a more democratic organization was seen as essential to its success. Although Minns described this shift as initiated by the Farm Bureau itself, the evidence suggests that the Grange played an equally important role in the formation of the new grassroots group.

Reflecting the priorities of farmers, the constitution of the FIA put this article first in its statement of purpose: "To foster cooperation in the buying and selling operations necessary to farming." "To assist in the operation and promote the usefulness of the Broome County Farm Bureau" came second.

Although the FIA supported the Farm Bureau and took over some of its educational functions, it placed those activities within a framework that was significantly different from that of the businessmen and agricultural experts. Cooperation in buying inputs and selling products would enable farmers to exert collective power in the marketplace, lowering their costs and raising the prices they received. Changes in farm practices were complemented by, rather than substituted for, concerted economic action. Pomona Grange Master Quinn was elected president of the new organization and "empowered to represent the association in any joint agreement entered into by the association respecting the farm bureau."[20] One of its first activities was to organize the cooperative purchase of lime. Lot J. Emerson of Maine served on the countywide committee.[21]

A year after its founding, the association met "for the purpose of deciding whether or not the Farm Improvement Association shall take over the work and responsibilities of the Farm Bureau."[22] Maurice C. Burritt, state director of the Farm Bureau,[23] said "that the Farm Bureau will do more good if conducted and supported by the farmers themselves, rather than the Chamber of Commerce & etc." Moreover, "it is doubtful if the state and government will support the work longer if the farmers are not willing themselves to assume the responsibility for the Farm Bureau." The agency had obtained public funds from the Broome County Board of Supervisors in 1912 and from New York State in 1913. Now that federal funds were becoming available under the Smith-Lever Act, it was even more important for the Farm Bureau to declare its independence from private interests. The motion carried unanimously.

Although the official history of the Farm Bureau describes this development as a Farm Bureau takeover of the FIA, both the minutes of the organization and the historical pageant put on by farmers in 1921 to celebrate "the 10th anniversary of the Farm Bureau idea" declared that this year-long process amounted by the takeover of the Farm Bureau by farmers themselves.[24] The nascent perception that agricultural improvement was a public responsibility, rather than something to be left to private interests, was helpful as farmers cast themselves as democratic representatives who ensured that public interests—as farmers themselves defined them—were to be served. Significantly, even though the Farm Bureau now existed across the state, county farmers did not change the name of their own organization; rather, they continued to meet as the independent Farm Improvement Association of Broome County and used the Farm Bureau name for the agent who was supported with public funds and for the activities they authorized him to conduct.

Cooperation was foremost on the agenda of Broome County farmers. In 1914 and 1915, they invited speakers to address the practical details of cooperative purchasing and marketing. At the very first meeting, Charles C. Mitchell of the Dutchess County Cooperative Association "spoke of the savings to farmers by the cooperative purchasing of seed, fertilizers, and supplies, and a Cornell professor spoke on the establishment of a public market in the city."[25] E. R. Minns, who served as county agent from 1913 to 1915, was a strong proponent of cooperation. His 1914 circular, "Broome County: An Account of Its Agriculture and of Its Farm Bureau," states, "The farm bureau agent is ready to help in the formation of clubs of farmers for buying supplies, for selling produce, and for assisting one another in social and educational ways."[26] Minns lacked his predecessor's rapport with local farmers, however; at the end of 1915, the Farm Bureau fired him.[27]

The Scientific and Business Model of Agricultural Progress

From 1914 through 1916, the Farm Bureau program included a variety of activities beyond promoting cooperation. The Western Broome County Cow Testing Association and the Holstein Breeders' Association were formed and conducted their business through the larger group. Testing the butterfat content of milk assisted farmers who sold to creameries and introducing new breeds that yielded larger volumes of milk was important to those who sold fluid milk. The Farm Bureau sponsored "agricultural contest-work" with school children and at agricultural fairs, pioneering the approach later developed by 4-H. The county agent was busy. In his 1914 annual report, Minns noted that "310 farmers had been personally visited on their farms" and "in his office the agent has answered 125 calls from farmers and persons directly interesting in farming." He had also addressed "fifteen farmers' gatherings and ten other meetings." He disseminated information about agricultural methods by means of letters to farmers, articles placed in local newspapers, and a circulating library. Demonstrations focused on liming the land, improving pastures and meadows, cultivating new types of hay and feed crops, renovating orchards, and cultivating potatoes.[28]

Minns was an enthusiastic advocate of the application of business methods to farming. He distributed 250 record books to promote the keeping of accounts, encouraged dairy farmers to keep production records, and personally conducted "a farm survey of two communities." He analyzed 37 records to compute farm labor incomes (the returns on farmers' investment of their labor and capital in specific operations) and publicized the results. He also cooperated with E. G. Misner of the New York State College of Agriculture on the first survey of dairy farming in Broome County.[29]

Minns's 1914 pamphlet expresses the experts' point of view on "the agricultural conditions and possibilities of the county." After analyzing prevailing problems, he went on to "point out desirable systems of farm management [and] suggest changes which should be made."[30] His list of the chief advantages of the county reads like an enumeration of its main difficulties: "Good farm land may be had at low prices." The availability of cheap land was the result of declines in both the rural population and the number of farms, but to Minns one family's misfortune was another's opportunity—provided they had "sufficient capital to invest." "During the decade 1900–1910, there was a shrinkage of 393 in the total number of farms owing to the inevitable and desirable consolidation of certain small farms and to the absorption of others by the growth of manufacturing industries." In his eyes, farm consolidation was a healthy development because the larger enterprises that survived were more efficient and productive than the marginal farms that had succumbed. Farmers could expand the scale of their operations by buying or renting the land vacated by their neighbors, and "investments pay reasonable profits." It did not occur to Minns that some families did not regard their farms merely as a business venture. In his view, those who lacked the capital required for profitable farming were natural casualties of progress.[31]

Minns's "suggestions for the improvement of Broome County farming" focused on taking advantage of growing markets for vegetables and poultry products in the Triple Cities and for milk in New York City. He advised dairy farmers to diversify their operations: "Dependence on the monthly milk check as the sole source of income is usually not so profitable as the combined sale of milk and that of cash crops." Herd improvement was a pressing necessity; dairy cows that were "unprofitable" should be eliminated, and purebred stock would "increase the average net return on each cow." The conservation of pastures and the improvement of meadows were equally important. "Especially on the hill lands…it is next to impossible to obtain satisfactory crops of clover and timothy without the use of lime." He recommended laying drainage tile because clay soils often retained too much water.[32]

Turning to farm management, Minns complained that this subject "has not received adequate attention." In his view, farm practices could be properly evaluated only when record-keeping allowed their profitability to be calculated. Minns deplored the pervasive neglect of "farm bookkeeping." Farmers' ignorance of business methods extended to other aspects of their operations as well. They seemed to improvise rather than systematically plan their activities, failing to "lay out their fields so as to economize labor," to adapt their crops to specific soils, and to practice scientific crop rotation. Business methods were crucial in enabling farmers to lower their production costs and thereby raise their profits. Even cooperation in the buying and selling of farm products appears, in Minns's account, as simply an advantageous way of doing business.[33]

Minns concluded his evaluation of Broome County agriculture with this declaration: "On too many farms in the county the total amount of business done is too small to provide a satisfactory labor income to the farmers. It is, therefore, desirable to increase the total amount of business done if possible. This may be accomplished…by increasing the intensity of operation (if the nature of the crops grown makes this profitable) or by renting additional land."[34] Enlarging the poultry flock, market garden, or orchard and improving meadows and pastures so that a farm could support more cows would require additional labor and more working capital. He suggested that dairy farmers in the valleys buy up hill farms to use as summer pasturage, assuming quite correctly that the larger and more prosperous farms would have to expand to remain profitable. Conversely, he took for granted that farms that lacked sufficient capital to expand would disappear. The model of farming as a business enterprise pervades his analysis. Although he recognized that the adoption of business principles would transform rural society, Minns never questioned the merits of that change.

Joining the Dairymen's League

The subsequent actions of the organized farmers make clear that their disagreements with Minns involved not only his personality but, more important, his model of agricultural improvement, which focused on cutting production costs by expanding the scale and increasing the intensity of farm operations rather than on raising milk prices. The solutions he and other agricultural experts recommended were too costly for most farm families to adopt. Equally seriously, they failed to address farmers' lack of power in the market, which many saw as the root of their difficulties. Organized farmers attempted to raise their incomes by demanding higher prices for their products. In their hands, producers' cooperation became a tool for restructuring the capitalist market.

Although the official records of the Broome County Farm Bureau do not mention the Dairymen's League before January 1917, local farmers were involved in collective action conducted under its auspices by fall 1916. Founded in Orange Country, New York, in 1907 by farmers who shipped their milk to New York City, the organization bargained collectively with the milk dealers to raise the prices that farmers received. The Dairymen's League took root among Grange members and spread across the milk shed through the group's well-established communications network. It found a fertile field in the Nanticoke Valley. When the cooperative creameries in Maine and Union Center were taken over by a profit-making company based in Binghamton in early 1915, farmers were dissatisfied. Eldon and Louella Rozelle, who farmed near Maine village, organized nearby farmers to sell milk through the Dairymen's League instead.[35] On March 13, 1915, Ralph Young recorded, "Had a meeting at our house tonight to talk over the milk business: whether to sell at Union Center or Maine." They decided to take their milk to Maine village every morning, taking turns hauling everyone's milk cans two and a half miles north on the Nanticoke Creek Road.[36] Soon the Young family became involved in the Farm Bureau as well. In January 1916, George was elected to represent the Grange on the Farm Bureau Board of Directors.[37] Then he recruited the whole Union Center Grange. Farmers' determination to engage in collective action soon pushed the Farm Bureau in the direction of cooperation with the Dairymen's League.

In fall 1916, when the Dairymen's League was strong enough to call a strike encompassing the entire New York City milk shed, Eldon and Louella Rozelle were right in the thick of it. So were the Youngs. After attending a meeting of the Dairyman's League in Maine, they held a meeting of their own in Union Center. On September 30, Ralph wrote, "Separated our milk tonight as there is to be no milk sent in the morning. A milk strike of farmers to get better prices." The next day, "We separated milk for the Scherters and the Strattons. No milk went from around here to Maine." In an interview, Ralph recalled that farmers in the vicinity were divided: "Our family struck, and the neighbors got down on us because our family was...trying to get better prices for the milk, and the rest of them wanted to take the price they could get...in their hand." There "weren't any physical fights; there was just different actions on it. We were on friendly terms with everybody all the while." The Rozelles' son recalled that "there was dumping of milk. It was more or less like any labor strife. I don't think there was any violence around here, but there was in other places."[38] But the strike held. On Sunday, October 8, Ralph Young wrote, "We commenced sending our milk again. Kinney Bros. have given in and will pay the price demanded by the Dairyman's League. All the companies except Borden's and two or three others have given the price and bought milk through the League." This victory was as important for establishing collective bargaining in the marketing of dairy products as it was for raising the price that farmers received.

In Broome County, the Dairymen's League was organized by the grassroots activists who had taken over the Farm Bureau. Jasper F. Eastman, who replaced E. R. Minns as county agent, was the brother of Edward R. Eastman, the Dairymen's League organizer in Delaware County.[39] A meeting of farmers producing fluid milk was convened in Binghamton in January 1917 after being announced at the Farm Bureau meeting and publicized by the new county agent. The next month, the meeting approved an agreement between the Farm Bureau and the Dairymen's League that separated the economic focus and concerted action of the farmers' group from the organizational and educational work of the publicly supported agency.[40] Yet it also facilitated close cooperation between them. In Broome County, the Farm Bureau and the Dairymen's League held two joint meetings in February and March. At the same time, the Farm Bureau moved its office to the county courthouse. As they began to organize against the corporations that processed and marketed their products, farmers were more comfortable in a public space than in the central offices of the business elite.

The next Dairymen's League strike in which Broome County farmers participated took place in 1919. During the war, the prices paid for farm products rose, but the costs of farm inputs and consumer goods rose even faster; federal wage and price controls provided only a limited amount of stability for the duration. When the war ended, inflation continued almost unabated, but there was serious downward pressure on the prices of farm products. While workers in industry responded to the declining real value of their wages with a series of mass strikes, farmers took collective action to raise commodity prices. Eastman, the county agent, proposed that farmers establish cooperative butter- and cheese-making facilities at local milk stations in case of a strike; if farmers could process their own milk and obtain some income, they would be able to hold out longer than if the milk were wasted. In February 1919, the *Broome County Farm Bureau News* congratulated the Dairymen's League on its victory, editorializing that "the Farm Bureau was glad to be of service to dairymen throughout the County" by publishing information and "helping the men to dispose of their milk" during the strike. In August 1919, at a picnic the two organizations cosponsored with the Grange, three thousand farm people heard a stirring address by E. R. Eastman.[41] The next spring, the Farm Bureau joined the Dairymen's League and the Grange in establishing the Grange League Federation (GLF), a consumers' cooperative for the purchase of farm supplies.[42]

The close cooperation among farm organizations was rooted in the grassroots nature of all three groups. When their separate histories are studied from an institutional perspective, in terms of their formal programs and statewide leadership, they appear to be different in their aims and activities. Within Broome County, however, the archival record makes clear that they were closely connected. The Grange played a key role in the formation of the Farm Bureau,[43] and the Farm Bureau, in turn, provided a framework within which the Dairymen's League organized and reached enough farmers to make the milk strike a success. At the local level, the three groups had overlapping memberships and rotating leadership, and their activities were often indistinguishable. As farm people undertook concerted action to alter their relationship to the market, they organized on the basis of existing associations and customary modes of action. Regardless of which organization sponsored a particular event or proposed a strategy, members of local communities acted in unison. The culture of solidarity and democracy that characterized rural social relations assured the success of the Farm Bureau and the Dairymen's League.

Farm Women Organize

While Broome County farmers, represented by the men in their families, were organizing producers' and consumers' cooperatives through the Farm Bureau, rural women not only supported these formally all-male groups but also organized their own activities and claimed a distinct place within the movement. The Grange had long tradition of gender integration; the national organization supported woman's suffrage, encouraged participation by married couples, and reserved some offices for women. As Ralph Young explained in an interview, "That's where the Grange was different than some other organizations, because the women there have the same

chance, and they can hold all the offices, besides two or three were specifically for women. In most organizations, women and men are divided." In
Broome County, Grange members who joined the Farm Bureau brought
these inclusive practices with them. Women frequently participated in the
demonstrations held on family farms and the meetings held in local community centers. As in most rural organizations, they took a leading role in
organizing social events to recruit and retain members. But women had no
part in the formal organizational structure of the Farm Improvement Association; the county officers and committee members were all men. Some
women felt that these men did not adequately represent them and were
convinced that they needed to organize among themselves. They found support among farm men, who interpreted their position differently than most
middle-class urban men; they took women's presence into account in their
deliberations and recognized the importance of their work. So, too, these
men supported women's efforts to make a place for themselves within the
Farm Bureau.

During World War I, the need to maximize production and conserve
food led the federal government to institute programs promoting gardening and canning. Officials recognized that, despite recent trends toward
large-scale row cropping and industrial canning, these activities remained
largely in the hands of women on diversified farms.[44] The pressure canner,
developed initially for centralized plants, was adapted to household use and
became the centerpiece of these programs. Canning clubs were organized
so women could learn new techniques and share equipment. This campaign
dealt with rural women collectively, and cooperation was the order of the
day. In Broome County, this program was conducted largely within the
Farm Bureau.

Work among women might have been terminated at the end of the wartime emergency, but the annual Farm Bureau meeting held in December
1917 resolved "to continue the work of the Home Economics Department"
in 1918.[45] The Broome County Cooperative Extension archives contain a
typewritten speech, "A Home Economics Department for Broome County,"
which was probably given by Ruby Green Smith of the Extension Service of
the New York State College of Agriculture. It argues that home economics
is much broader in scope than feeding families, encompassing all the factors that contribute to farm people's well-being. Customary ways of doing
things, handed down from mothers to daughters, are not necessarily the
best; housework requires expert knowledge. Household tasks can be done
more efficiently if they are subjected to scientific study and systematization.
Aware that some women found housekeeping lessons from self-described
experts insulting, the author was careful to assure them that advisors would
not belittle their skills: "The work of the County Home Demonstration
Agent will not be to attempt to teach the women of the County how to
do work in their own homes which they already know how to do," but,

rather, to lecture and conduct demonstrations on specific subjects suggested by farm women. Home economists' advice would supplement, rather than substitute for, rural women's expertise.[46] This proposal was accepted by the organized women and men of Broome County, who authorized the creation of a Home Economics Department within the Farm Bureau.

The 1918 meeting of the FIA entertained a proposal from H. W. Babcock, state leader of Farm Bureaus, to adopt a new constitution renaming the organization the Broome County Farm and Home Bureau and inserting "home economics" alongside "agriculture" in the statement of purpose. This constitutional change had already been adopted by a number of Farm Bureaus across the state.[47] Initially the proposal was rejected by a vote of the men present, but a committee was set up to study the question. The next month, a special meeting was held to "to consider changing the name of the organization from the Broome County Farm Improvement Association to that of the Broome County Farm Bureau and the adoption of the Home Economics organization as a department of the Farm Bureau." This proposal essentially codified the situation at that time, recognizing that the FIA need not maintain a separate identity to ensure democratic control of the Broome County Farm Bureau and giving home economics a subordinate status within the bureau. But Julius E. Rogers, past president of the FIA, and Duane Barnes of Glen Aubrey, a resident of the town of Nanticoke who served on the countywide Board of Directors, moved "that the name of the organization be the Broome County Farm and Home Bureau." This proposal was favored by state leaders and by agricultural and home economics experts, and it had secured the support of key local activists.[48] Ruby Green Smith did not credit her own speech with persuading people to change their minds. Instead, she recounted, "One of the shortest and most eloquent speeches ever made in behalf of home life was made by a Broome County farmer. He rose and said: 'Mr. Chairman, Lady, and Gentlemen; Home is a beautiful word. The oftener we say it, the better. I move the adoption of the new name.'" The motion carried unanimously.[49]

Women gained an integral and recognized place within the Farm and Home Bureau Association of Broome County. Ruby Green Smith described the organization as "laying the foundations for a better system of farming methods and the lightening of labor for women on the farm."[50] The constitution and bylaws provided that the farm and home departments be co-equal. Members might belong to one or both departments; although the Home Bureau was intended specifically for women, women could also join the Farm Bureau, and joint membership by husbands and wives was encouraged. Each department was empowered to conduct its own activities, and a joint board of directors coordinated their programs and addressed common concerns.[51] The participation of women required the installation of a ladies' lavatory in the courthouse near the Farm and Home Bureau office, marking a significant advance in the accommodation of women in public space.[52]

These developments were shaped by the contemporary political context, as well as by farm families' preference for cooperation across gender lines. Women had won the suffrage in New York State in November 1917. An article published in the June 1, 1918, issue of *The Nation* reported that the New York Woman Suffrage Party had decided not to dissolve but to continue as a nonpartisan organization and conduct educational work to prepare women to exercise their citizenship rights wisely. The party listed mobilizing rural women alongside reaching immigrant women as among its most pressing priorities.[53] Equally important, the acceptance of the Home Department alongside the Farm Department took place immediately after the Dairymen's League strike. The active role that women played in that struggle may have prompted men to recognize women's rights to full and equal participation.

Nanticoke Valley Farmers' Activism

In 1919 and 1920, following the successful milk strike and the incorporation of the Home Department, the Farm Bureau organized groups in rural villages and open-country neighborhoods throughout Broome County. Most histories of the Farm Bureau describe it as a countywide rather than locally based organization and contrast its structure with that of the Home Bureau, which was constituted by active local units. The evidence from the Broome County archives suggests that this contrast is overdrawn, at least for this region. Even more significantly, the model of gender segregation that both agricultural and women's historians assume defined the Farm and Home bureaus is inapplicable to Broome County. The Farm Bureau was not a male-dominated, top-down organization, nor was the Home Department a ladies' auxiliary. At the local level, women and men organized and acted together, although women were also able to organize separately. In the Nanticoke Valley, where the grassroots leadership and activities of these organizations can be reconstructed in detail, democratic practices and gender integration went hand in hand. After "practical farmers" took control and the FIA made room for organized women, it became more deeply rooted in particular communities. Local organization and initiative were seen as essential for securing broad participation and ensuring the solidarity that collective action required.

By 1919, community meetings were regularly held in Glen Aubrey, Maine, and Union Center. Several men from the Nanticoke Valley had already been active at the county level.[54] In 1919 and 1920, the Farm Bureau conducted campaigns to increase its membership and strengthen its local units and elected "community Farm Bureau leaders." In the Nanticoke Valley, a dozen men were chosen in the first year, one from each school district in the towns of Maine and Nanticoke; the next year there were two dozen. Although the membership lists that survive in the county archives do not identify individuals by locality, at least four dozen names included on a 1919 list are identifiable as Nanticoke Valley residents.[55] Notably, this list contains the names of several women. Most are identified as "Mrs.," but others, such as Mary E. Kelley, are listed under their own names. Although the county-level leadership was drawn primarily from substantial farmers from long-established families, the membership was strikingly diverse, including farmers with relatively small-scale, diversified operations; those who lived in the hills; and newcomers. Impoverished Corsons and immigrant Tartanians appear alongside Mareans and Allens who had been in the locality for generations. The Farm Bureau enjoyed a wide base of support in the Nanticoke Valley; groups continued to meet in Glen Aubrey, Maine, and Union Center throughout the 1920s.

Hardly a month went by without a report from the Nanticoke Valley appearing in the *Broome County Farm Bureau News.* Farmers participated actively in field testing of new crops and techniques. In 1919, six Nanticoke Valley farms were featured in articles on improved agricultural practices: Louis Ketchum and Fitch Marean tested new varieties of ensilage corn; George, Stacey, and Ralph Young raised high-yield oats for cattle feed; and Venley McGregor began to irrigate vegetables. Meetings were held regularly during the late fall, winter, and early spring, and demonstrations were scheduled during the growing season. In January 1921, for example, the Maine group met to discuss the keeping of farm accounts, and the Union Center group studied cattle breeding. The next winter, sixteen members of the Maine group gathered at Eldon Rozelle's to consider dairy cows while the Glen Aubrey group met under the chairmanship of Ralph Carley to discuss farm economics.

Nanticoke Valley farmers organized programs on subjects of their own choosing, selecting the topics that were of greatest concern to them from the resources available through the Farm Bureau. Demonstrations conducted by the county agent and, occasionally, an expert from Cornell explained best practices for dairying, poultry raising, and market gardening in an applied setting. According to the *News,* poultry culling demonstrations were held one year at the McGregors, the next year at the Carleys, and the year after at the Woodwards. Photographs of those events show groups of men, women, and young people gathered in the farmyard watching as a man in overalls talks about the hen he is holding. Ralph Young's diary records that in 1921 he and Stacey visited three orchards, including the Green Brothers in East Maine. The next year, a group visited the Youngs: "There was a FB [Farm Bureau] Farm Management Tour around different farms in the Co. They came to our place just before noon, were here about 3/4 of an hour. Father, Mother, Ella and children went on with them." These sessions reached not only the farmers who attended them but also the neighbors whom they told what they had seen and heard. Reports of experiments published in the *News,* too, amounted to invitations to neighbors to drop by and inspect the operation. Farm Bureau activities built on the informal modes of work sharing and visiting that circulated information and ideas throughout the community.

Men and women from the Nanticoke Valley served as county Farm and Home Bureau leaders throughout the 1920s. Significantly, those who did so had their base in local organizations, especially the Union Center Grange. George Young served on the board almost continuously and represented Broome County at the American Farm Bureau Federation convention in Chicago in 1924; his wife, Alice, accompanied him to these meetings. Young played musical chairs with Harry Woodward, who also served on the Pomona Grange and the Farm and Home Bureau board.[56] While the men held elective office in the Farm Bureau, their wives either led the Home Bureau or shared the men's duties. According to the Rozelles' son, Louella conducted the correspondence and kept the books while Eldon was county director.[57]

The Farm Bureau undertook public advocacy at the state and national levels throughout the 1920s, campaigning to encourage the consumption of butter rather than margarine and lobbying for federal programs such as the rural free delivery of mail. At the same time, Broome County farmers continued to take concerted action to exert control over the market for their products. At the county and local levels, they used the Farm Bureau to facilitate the work of their cooperatives. The close connections among the Grange, the Farm Bureau, and the Dairymen's League are evident in the Nanticoke Valley. Edward F. Vincent of Maine was elected vice-president and then president of the countywide Dairymen's League in 1919 and 1920. Vincent's name appears regularly in the *Broome County Farm Bureau News*, not only in his official capacity but also as a Farm Bureau member who joined the Cow Testing Association and participated in the New York State reforestation program. Edgar L. Vincent contributed a column on events "among the farmers" in the Southern Tier to E. R. Eastman's *American Agriculturist*.[58] In Maine and Union Center, Dairymen's League branches and Farm Bureau locals held joint meetings.

The organization of the Grange League Federation (GLF), a consumers' cooperative, exemplifies the working relationship among these farm organizations. Incorporated in June 1920, the GLF conducted a membership campaign in July. George Young served as Broome County chairman. In Maine, Lot Emerson of the Farm Bureau, Edward Vincent of the Dairymen's League, and Jerry Wright served as chairmen. Fred Holden, who became a local Farm Bureau leader later that year, led the drive in Nanticoke; Wayne Woodward of the Grange led the Union Center campaign. Within a month, there were 74 members in Maine and 37 in Nanticoke. By the end of 1922, the Tymeson brothers, who operated a store in Maine village, were taking GLF orders from farmers and Harry Woodward was serving as GLF agent in Union Center. Late that summer, Glen Aubrey hosted the annual Broome

County Farmers' picnic, cosponsored by the Farm and Home Bureau, the Grange, and the Dairymen's League.[59]

In 1920 and 1921, the *Farm Bureau News* reported that the Dairymen's League branches in Maine and Union Center were among the first to incorporate as cooperative associations. In October 1920, the fluid milk dealers of Binghamton agreed to the Dairymen's League's proposals about prices, but the condensed milk dealers did not, and all the manufacturing plants except Kinney Brothers of Binghamton, which operated the plant in Maine village, closed their doors. This lockout left those Dairymen's League members in Union Center who did not ship fluid milk without a local outlet. The *News* stated that "some branches have facilities for making butter and cheese. In places where this is not possible, farmers are making the milk up at home as best they can." To deal with this situation, which might have pitted farmers with smaller dairies against those with larger herds, the Dairymen's League adopted a "pooling plan," providing that farmers who sold fluid milk would share the risks of those who sold to condensaries. The *News* reported that the Dairymen's League was making slow progress in getting pooling contracts signed because fluid milk producers did not necessarily sympathize with the smaller and more remote producers outside the New York City milk shed. But Nanticoke Valley farmers readily signed pooling contracts. By February 1921, three months before pooling began, 106 farms in Maine and 37 in Union Center had signed up. In November 1921, the *News* pointed out that the recent strike of the New York City milk drivers had demonstrated the success of the pooling plan because Crowley's of Binghamton had bought milk from Dairymen's League members.[60] The Youngs played an active role in the boycott of dealers who did not settle with the drivers, and after the settlement Ralph "was elected as a delegate to D.L. meeting at Jersey City" and traveled to the conference with five other Broome County representatives, two of whom were also from the Nanticoke Valley.

The incorporation of immigrant newcomers into the Dairymen's League was essential to the success of the pooling plan. When economic cooperation was concerned, Nanticoke Valley farmers openly criticized the anti-immigrant prejudice that animated such groups the Ku Klux Klan (KKK).[61] Families who were active in farm organizations made deliberate efforts to include immigrants in political-economic as well as recreational gatherings. In 1921, an editorial in the *American Agriculturist* entitled "Americanization" exhorted farmers to welcome "newcomers, immigrants and aliens" to their neighborhood, sharing knowledge, labor, and sociability rather than being "standoffish." Praising immigrants as hardworking, vigorous, and hardy, it suggested that they be invited to meetings and encouraged to participate in democratically run community organizations, especially to speak up and be heard, and assured readers that immigrants were ready to contribute to local groups and would reciprocate whatever welcoming gestures were made to them.[62] The political point was stated even more sharply in 1923, when an editorial denounced the KKK as inconsistent with American principles. Despite its claims to be defending Americanism, the editorial argues, the KKK uses vigilante methods and is prejudiced against members of a religious sect—Catholics—rather than upholding religious toleration.[63] The inclusion of those whom the Youngs called "new Americans" was central to the Dairymen's League's success at this crucial moment.

The close connections that the Farm Bureau maintained with the Dairymen's League led to changes in its structure and financial support during the early 1920s. In 1921, just after the celebration of the tenth anniversary of the "Farm Bureau idea," the DL&W notified the Farm Bureau that it was discontinuing its financial contribution on the grounds that the agency was now securely established and had other sources of support.[64] The break with the Binghamton Chamber of Commerce came two years later, at the instigation of the Farm Bureau itself.[65] The Farm Bureau was now entirely controlled by farmers.

A grassroots perspective on collective action is presented in E. R. Eastman's 1925 novel *The Trouble Maker,* a lightly fictionalized account of the milk strike in Speedsville, a hamlet just east of the Nanticoke Valley. The plot turns on the conflicts that arise within the rural community over participating in the Dairymen's League action. When a fistfight breaks out between the organizer, Jim, and a farmer who opposes the strike, many people are disturbed and blame Jim; even Dorothy, the young woman he is courting, condemns him as a "trouble maker." Tellingly, Jim is blamed for causing hostilities among neighbors when he organizes farmers to strike, but is vindicated when he shows that collective action is consistent with local customs of mutual aid among farm people. The novel culminates in the victory of the strike and the wedding of Dorothy and Jim. Harmony is restored, but on a new, less individualistic economic basis.

The most dramatic scenes focus on the women's attitude toward the movement. When Dorothy is talking with Jim about whether farmers will stick together and stand firm against the dealers, he argues earnestly that raising milk prices will benefit women as well as men. Reminding her that his widowed mother was worn out by a lifetime of labor, he generalizes: "Farm women have been worked to death, slaving in the house without any conveniences, taking care of large families, and then like as not, having to milk some six to a dozen cows every night and morning." Then he comes to the crux of the matter between them: "I know that unless conditions change, I can't ever ask the girl I love to share them with me." Without acknowledging his avowal that he would propose to her if he could, she responds, "Yes, but it is no worse for the women than it is for the men. One of the nice things about the farm business is that the women-folks are real partners with the men, and most farm girls want to share the burden."[66]

10

Home Economics and Farm Family Economies

The development of the Home Bureau, like that of the Farm Bureau, was marked by clear differences between the plans of outside experts and the practices of rural residents. The reform proposals made by professional home economists were dissonant with farm women's perceptions of their situation and definitions of their interest. But in the Home Bureau, in contrast to the Farm Bureau, these differences did not lead to open conflict within the organization. Right from the start, the Home Bureau was democratic and decentralized. Its structure was more in consonance with rural women's customary modes of collective action than that of the Farm Bureau, and women did not have to seize control of the organization from outsiders. When faced with ideas about farm family economies they did not agree with, they simply ignored the experts. They adopted only what they found useful from a long menu of programs.

Most notably, farm women rejected the model of gender that was implicit in the dual structure of these organizations and explicit in much of the expert advice they received. The notion that women and men did, or should, occupy "separate spheres" was profoundly foreign to their experience and understanding of gender relations.[1] Women and, to a striking degree, men shared a perspective on family farming that emphasized cooperation and reciprocity rather than differences and divisions, even when gender distinctions were presented as complementary rather than restrictive. As Broome County activists united in the Farm and Home Bureau, women continually crossed the hypothetical boundary between farm and home concerns.

Farm women used the Home Bureau to claim a legitimate place within the county and statewide organization and gain access to its resources, rather than allowing themselves to be excluded or marginalized. At the same time, well aware of the tendency of large-scale organizations to become male-dominated, they promoted more local forms of organizing and involved men in a wide range of community-centered, gender-integrated activities. In the Nanticoke Valley, men cooperated with women's efforts to sustain more equitable modes of labor, decision making, and collective action.

Home Bureau Programs for Farm Women

Home economics professionals, unlike women in the Nanticoke Valley, shared Mary Wilkins Freeman's assumption that farm women's problems stemmed from putting the barn before the house. Deeply committed to improving the condition of rural women and raising their status in society, they demarcated a distinct domestic domain within which women could exercise power and which, they believed, should be recognized as equal in importance to the supposedly masculine domain of farming.[2] Both the structure of the organizations that home economists established and the practices that they advocated presumed a gendered separation between home and farm.[3] This approach often put them at odds with farm women. Rural women refused to cede responsibility for the farm operation and authority over major financial decisions to men. They upheld and sought to sustain family farming as a partnership based on shared labor, joint decision making, and common participation in social and political activities. While adopting those Home Bureau programs that enhanced family and community life, they emphasized that earning money was a higher priority than mere homemaking.[4] Broome County women concurred with their sisters across the nation who responded to the question, "What do farm women want?" posed in 1926 by the American Country Life Association and the *Farmer's Wife* magazine: "The one want most often recorded...was money, more money....Many said that more money would solve their troubles."[5]

The Home Bureau's objectives, organizational structure, and program offerings were designed by home economics professionals.[6] Ruby Green Smith of the Extension Service of the New York State College of Agriculture played a leading role, both through her supervision of the first home demonstration agent and through her influence on Farm Bureau leaders. When the Home Department was granted coequal status by the Broome County Farm Improvement Association in 1919, its basic structure was already in place. The home demonstration agent, Alice Ambler, was working with ongoing groups of women in rural communities. The programs offered during the 1920s extended far beyond food conservation and canning to encompass nutrition, health, and home nursing; sewing, making over, and mending clothing, or "the art of making do"; household efficiency, based on "Bulletins for the Farm Home" published by the Home Economics Department at Cornell,[7] which advised women how to organize their tasks so they required less time and energy; child welfare, which included the public provision of school lunches; and, after the advent of woman suffrage, "the study of civil government."[8] Concern for the welfare of the rural community is especially prominent. The Home Bureau section of the *Broome County Farm and Home Bureau News* stated that "the best type of Home Bureau Club works through its members for the general betterment of the community, rather than merely for the good of its individual members."[9] Jennie C. Jones, assistant state Home Bureau leader, emphasized that the organization should help "the less favored women of the community" and children in rural schools,[10] which were the subject of a recent statewide investigation and report. The Home Bureau suggested projects to improve school programs and facilities, such as the purchase of instructional and

Figure 11. "Judging canned goods at food exhibition, Bath Fair," New York, September 1945. *(Photograph by Charlotte Brooks, Standard Oil (New Jersey) collection, University of Louisville Photographic Archives. [SONJ 30380])*

playground equipment. Other community projects included "women's ex-changes," where rural women could sell their handiwork directly to con-sumers; "community closets," which loaned out supplies for the care of the sick; and recreational programs, such as "community sings," that included women and men, adults and children.[11]

Home economics professionals accepted the domestic definition of wom-en's place that prevailed both in the dominant, middle-class urban culture and in agribusiness and sought to restrict women's participation in farm labor in an effort to raise their status as homemakers. At the same time, they enlarged the boundaries of conventional feminine domesticity to in-clude significant public activity. As Martha Van Rensselaer, co-chair of the Home Economics Department at the College of Agriculture, defined it, "The purpose of home economics...is to aid directly in solving the problems of home and family life, and to extend this aid to the solution of such of those problems of community life as are extensions of home activities."[12] The no-tion of "social housekeeping" recognized that the quality of community life, especially health, education, and welfare, impinged on the family at home and claimed that women's public activities were a natural extension of their domestic responsibilities. This rationale, which was articulated by feminists during the Progressive era, justified the active involvement of women in edu-cational, cultural, and political affairs.[13] A distinctly rural version of social housekeeping is evident in Home Bureau programs. Although the organiza-tion focused on women's domestic labor, it recognized that rural women's responsibilities for children, health, and charity had always led them into community work. Home Bureau programs promoted neighborly sharing of expertise and resources, and the group endorsed and extended women's roles as community organizers.

The Home Bureau was democratic and decentralized. Local groups had membership committees composed of one woman in each school district, a system for recruiting members later adopted by the Farm Bureau. Schools serving open-country neighborhoods were the most inclusive rural social or-ganizations, as well as an important focus of women's civic activism. Home Bureau groups should be welcoming to all: "Let us keep as far away as possible from the exclusive club idea."[14] Equally important, each Home Bu-reau unit chose its own programs from those offered in the county during the year. Community groups exercised even more power after 1925–1926, when the organization shifted to a local leader system; after attending one-day training programs on specific projects, members passed on the informa-tion and skills they had learned to their peers. This structure helped women develop their leadership capacities and brought women from different lo-calities together in an informal setting.

In 1923, Anne Phillips Duncan became the home demonstration agent for Broome County. Her detailed plans of annual offerings assume an en-tirely domestic definition of women's work. Neither dairying nor poultry raising was listed among the possible subjects of study. Instead, household production for family use was at the center of women's responsibilities. Improving nutrition and saving cash were the stated rationales for gardening, ignoring the fact that an increasing number of rural women sold vegetables in town.[15] The craft skills the Home Bureau taught, such as chair caning and rug hooking, were utilitarian as well as decorative, yet they might become sources of income.[16] The 1925–1926 program plan stated that fancy work "has been the means of making money by sale of articles" and recommended the establishment of a "Home Bureau exchange, to which rural men and women may bring in for sale the things they make."[17]

Projects on household management were intended "to lighten the burdens of housework, to raise the standards of housekeeping," and "to increase the numbers using and installing labor-saving equipment and devices." Home economists aimed to reduce the amount of "drudgery" that farm women performed, but their proposals amounted to a reallocation of women's labor rather than an overall saving of time.[18] Some programs aimed at lifting women's minds and spirits above the dull and confining routine of household chores and ensuring that they had healthful exercise in the open air. The 1925–1926 plan went so far as to say that "every women should have an outdoor hobby."[19] This declaration sounded strange to farm women, whose outdoor work represented not a leisure activity but a necessary contribution to the family economy. Few rural women complained of being stuck indoors all day; most went outside in all weather to fetch water, get the cows, do barn chores, and toil in the fields.

Other programs instructed women in new technologies.[20] Countywide automobile schools taught women not only how to drive but also how to perform routine maintenance and emergency repairs, which enabled them to run errands and attend meetings independently from men. In the 1920s, the image of a woman at the wheel or changing a flat tire without male assistance epitomized the competence and self-reliance of the "modern" woman.[21] As mobility increased, the Home Bureau tried to ensure that rural women were not left behind.

In order "to emphasize the importance of the responsibilities of homemakers," the Home Bureau taught women to keep household accounts and make budgets. Evidently the agent assumed that women did not adequately appreciate their own centrality to the farm family economy and lacked the skills necessary to manage financial matters. It is more likely that most farm women were well aware of the value of their work but realized that some farm men did not sufficiently recognize its importance, in part because women's independent incomes, exchanges, and purchases were not integrated into the farm accounts. Even without account books, women kept track of their own transactions, many of which turned on credit rather than cash. Still, the notion that only "petty" sums were involved may have reinforced women's perceived marginality. Home economists' lessons on financial management were double-edged. On the one hand, they assumed a clear separation between household and farm accounts; on the other, they attempted to prepare women to act independently and knowledgably in financial affairs. Rural women were likely to use this knowledge to participate in farm decision making rather than to make household consumption budgets as urban women might do. In a farming community, teaching account keeping as a way to emphasize the importance of homemaking is full of ironies. By treating the expenditure of money for the purchase of consumer goods as central to rural women's responsibilities, it accepted the devaluation of their continuing production for use—a "saving" that cannot be precisely measured in monetary terms but is an essential complement to wise "spending."[22] Nevertheless, rural women still thought of themselves as producers rather than consumers and as savers rather than spenders.

Home Bureau programs intended to equip women with the skills necessary to control their own financial affairs also covered "banking, insurance, and wills."[23] Farm women's fortunes often depended on the terms of deeds, mortgages, and contracts; they were equally concerned with the intergenerational transmission of the farm, which depended on the protection and transfer of property. Home Bureau programs focused more directly on preparing for the possibility of widowhood than on protecting the interests of women who were active partners in the farm enterprise, but the skills women learned were also applicable to farm management.

Community-oriented projects focused on child welfare, rural schools, and group recreation. The Home Bureau sought "to arouse interest in community halls, rest rooms, cooperative centers, etc."[24] "Rest rooms" were places where women who were in the city to sell their produce or purchase goods for their families could eat the lunches they brought, tend to their small children, use clean lavatories, and rest before starting home. Rural women were seldom comfortable using the ladies' lounges in department stores they did not patronize, and few public facilities were available. In "cooperative centers" women could share expensive equipment, such as laundry machinery. "Community enterprises" raised money for parks and recreational facilities. "Home talent plays," which used scripts written for amateur actors that were set in rural communities, became especially popular. Concern that the commercial amusements available in cities might exert a centrifugal force on local society prompted these efforts to sustain participatory, inclusive, and distinctly rural forms of social life.

A report on "Community Life and the Home Bureaus" written by Ruby Green Smith in the late 1920s describes numerous projects undertaken by local groups that promoted cooperation among women to meet vital social needs.[25] Smith's more radical proposals, such as "cooperative care of children," "cooperative ownership of expensive household equipment," and "exchange of cooked food, millinery, and clothing in community groups"[26] were not implemented in the Nanticoke Valley. But these suggestions represented the formalization of time-honored practices of mutual aid among rural women and of cooperative purchasing among farmers. As women on scattered farms discussed their common problems, they developed a sense of connection that spilled over into their daily lives, and friendships formed at Home Bureau meetings led women to share many tasks informally. Some of the projects Smith suggested were implemented at the county level. The Home Bureau set up day nurseries and playgrounds at its own meetings, the Broome County Fair, and the annual Farmers' Picnic. Through their activities as community organizers, women simultaneously improved their own lives and enhanced the quality of life for others.

Organizing the Home Bureau at the Grassroots

The women who founded local Home Bureau groups were, by and large, the wives of men who were active in the Farm Bureau. In their choice of programs, they emphasized not homemaking skills but, rather, ways of contributing to the farm family economy.

Acting in concert with the Farm Bureau, the Home Bureau conducted countywide membership campaigns in 1919 and 1920. The Maine group won the 1919 contest by doubling its numbers to 22.[27] The Glen Aubrey group, led by "Mrs. Duane Barnes" and "Mrs. Nathan Barnes"—whose husbands were leaders in the countywide and local Farm Bureau groups, respectively—doubled its numbers in 1920 and had 25 members by 1921. A meeting was held in the hamlet of Nanticoke, but no group seems to have been established there; the same was true of the Farm Bureau.[28] In Union Center, membership increased markedly in 1921 when women affiliated with the Grange joined the Home Bureau in a body at the same time that the men joined the Farm Bureau. By 1923, Union Center had 31 members and Maine had 29.[29]

The Home Bureau, even more deliberately than the Farm Bureau, presented itself as representing the interests of all women and dedicated to overcoming divisions among rural residents. Its membership was broad-based, encompassing women who lived in open-country neighborhoods as well as villages. Most Home Bureau leaders, like most Farm Bureau leaders, came from better-off, long-established faming families. But many members had small-scale farm operations and others only a kitchen garden. In demographic terms, the group was more homogeneous.[30] Almost all members were married. News items customarily identified women by their husband's name. Occasionally a married woman's first name would appear, and this practice became more frequent over time. Unmarried daughters seldom belonged, although they might participate in local activities, but single women, such as Mildred Whittaker of Glen Aubrey, joined in their own right. Most local leaders were older, and few had young children. In these respects, Home Bureau members resembled women who belonged to other community organizations.[31]

Indeed, the primary characteristic that distinguishes Home Bureau activists from women who were not members is their high level of participation in other community groups. Although the Home Bureau was in principle secular and nondenominational, in practice it was Protestant. During the 1920s, not one woman mentioned in Home Bureau news items was affiliated with the Catholic Church. Foreign-born women were almost entirely absent. Native-born daughters of immigrants who had grown up in the locality and participated in the social activities of local Protestant churches were members, but those whose families had recently moved to the Nanticoke Valley were not. This pattern was the natural outcome of the informal process of recruiting Home Bureau members. Women were brought in through existing networks of kinship, friendship, and affiliation. Women's organizations, such as the Ladies' Aid and Women's Home and Foreign Missionary societies, were especially fruitful sources of new members; so were gender-integrated organizations such as the Grange. Home Bureau leaders thought of their organization as inclusive, and in a socioeconomic sense it was. But the Home Bureau, in contrast to the Farm Bureau and the Dairymen's League, did not succeed in incorporating people who had previously been unorganized. Deliberate, sustained efforts were required to reach out to newcomers and persuade them to join in collective economic action, but they were assured of a welcome by farmers' organizations. It was much more difficult to bridge the cultural gap between native-born and immigrant women, many of whom did not speak English fluently. Native-born women who rarely attended church, seldom participated in social events, and did not visit the villages were also left out. Local leaders were well aware of these women; the Glen Aubrey, Maine, and Union Center groups aided "needy families" who came to their attention, and "work with foreigners" was explicitly mentioned in the countywide program during the late 1920s. But women in poor and immigrant families remained the objects of Home Bureau ministrations rather than becoming participants.

Many Home Bureau activists were the wives of Farm Bureau activists. An item in the *Broome County Farm Bureau News* light-heartedly announced that "Judson H. Sherwood forsook the life of single freedom for the yoke of double blessedness. We welcome his wife into the Association."[32] The writer assumed that Mrs. Sherwood would adopt her husband's organizational affiliation just as she would take his surname. Husbands and wives worked together in the Farm and Home Bureau in a relatively egalitarian way. The group's dual structure meant that women were not subsumed by their husbands, as could occur in some gender-mixed organizations, but had an organizational identity of their own. At the same time, their coordinate relationship meant that couples could study farm and rural problems, attend countywide meetings, and engage in community projects together. Those who did so tended to be elected to leadership positions in both organizations. Delia Whittaker Barnes of Glen Aubrey served on the Broome County Board of Directors throughout the 1920s while her husband, Duane, held local and county-level offices. Other Nanticoke Valley women who served as Home Bureau directors and committee chairs were married to Farm Bureau leaders as well.[33] In each locality, the core of the Farm and Home Bureau was composed of extended kin groups: in the town of Nanticoke, the Gaylord, Kenyon, Riley, Walters, and Whittaker families; in Maine, the Bowers, Marean, McGregor, and Vincent families; and in Union Center, the DuBois, Ketchum, Smith, Westcott, Woodward, and Young families. Women and men worked together in these coordinate organizations as they did on their farms, and they conducted their organizational activities, as they did their social lives, within a dense gender-integrated network of relatives, neighbors, and friends.

Nanticoke Valley Women's Priorities

The projects undertaken by the Glen Aubrey, Maine, and Union Center groups emphasized labor-saving, cash-saving, and income-producing activities. For example, the Maine group considered the "fireless cooker," which was designed to save labor and fuel by heating food and then letting it cook in an insulated container.[34] In Glen Aubrey, cooperative buying of pressure canners and fireless cookers was organized by Delia Whittaker Barnes and her daughter-in-law.[35] Doing routine household chores more efficiently enabled a woman to devote more time to her poultry, the dairy, and other types of farm work. Other programs aimed to strengthen women's position in the family and in public life. "Business Methods for Women," based on a Cornell bulletin, promoted women's financial independence and prepared them to participate more fully in farm decision making. "Keeping Fit" included information about venereal disease, which had become a matter of concern during the war.[36] Women in the Nanticoke Valley regarded the Home Bureau not only as a resource for advice about how to perform their customary tasks more efficiently but also as a means of gaining insight into political, economic, and social questions.

Home Bureau groups extended women's roles in organizing community events. Nanticoke Valley women reshaped Home Bureau programs in accordance with their gender-integrated and inclusive modes of sociability. Adapting organizational forms to local customs, they erased the boundaries between formally separate organizations and acted through several groups simultaneously to achieve the goals they defined for themselves. Home Bureau activities resembled those of the Ladies' Aid societies based in the Protestant churches. Although both were designed as female auxiliaries to formally male-dominated organizations, in practice they facilitated women's participation in public affairs and organized activities for men as well as women and children.[37] In many cases, these groups worked in concert.

The Home Bureau groups in the Nanticoke Valley express the distinctive interests and orientation of women at the grassroots. As historians have observed in other regions of the United States, many farm women did not accept home economists' diagnoses of the problems they faced or their prescriptions for curing them. Instead, they held on to a less gendered perspective that focused primarily on farm economics. As they participated in extension programs for women, they took up experts' suggestions very selectively and adapted them to their own circumstances. Most often, they focused their energies primarily on ways of making money through work they could do at home and on saving time in doing essential domestic chores so they could spend more time working on the farm. Contributing to the family income was their central goal.[38]

Farm women across the region shared these views. In 1924, E. R. Eastman, editor of the *American Agriculturist*, argued in "Should Women Help with Farm Work? Their Sacrifices, Sometimes Necessary, Have Not Always Paid" that farm women were overburdened by labor and ought not to be expected to work outdoors. After recalling that his mother milked and raked hay, he stated that women have done their share, "or perhaps a little more than their share, and certainly more than their health would have permitted, of the heavier work in the barn and fields." Now, with young folks gone and hired labor scarce, women are doing more: "The work of the woman has made it possible on thousands of farms to continue the business." Mentioning native-born women toiling on hill farms and immigrant women raising vegetables for the market, he acknowledged that peasant women are strong but lamented that they are treated as "beasts of burden." The piece concludes (illogically) that "the cheaper free labor of women folks of the farm has been one reason for the low prices of farm products." In his opinion, outdoor work should not be added on to housework; those women who must do farm work cannot be expected to do housework at the same time.[39]

Although Eastman's view was not as categorical as the gender separation espoused by agricultural professionals, the editorial elicited a storm of criticism. The *American Agriculturist* published a selection of "Letters from Our Women Readers: They Say Outdoor Work Does Not Make Women Drudges." Mrs. E. L. declared that she helped her husband outdoors voluntarily and reported that educated young women said that "with more household conveniences they'll have more time to take a part in outdoor work and not neglect their house either. They're going to be their husband's partners even more than before." Mrs. R. C. stated that with household conveniences to lighten her indoor labor, she could do more farm work; she was proud that she had raked hay at the end of the day, even "by moonlight," "milked my part of the cows," and made $5 a day from the hens.[40] When the farm paper sponsored an essay contest on the question, "Which—Machinery for the Farm or the Home?" those who responded did not endorse the views of home economists; instead, a woman said that the first priority should be "farm equipment needed for profit" so as to avoid hiring labor, and a man said that the home and farm are "a cooperative business" and both need modern equipment.[41]

Throughout the 1920s, Home Bureau meetings held in the Nanticoke Valley focused on cash-saving and income-producing work and on community organizing. In the most popular projects, women conducted social events and devoted the proceeds to charitable purposes. They put on suppers, bake sales, and ice cream socials to raise funds for schools, village greens, cemeteries, and community halls. In keeping with rural culture, women planned social occasions for their whole families. The Maine group presented a home talent play, "Mrs. Briggs of the Poultry Farm," to an audience of one hundred at the community hall.[42] The Glen Aubrey group put on "The Old Green Plush Album" at the annual Farm and Home Bureau meeting in Binghamton in 1927. The premise of this play was simple; the portraits of four generations of an extended family in a photograph album come to life. The playbill lists eighteen characters, including not only a cast of cousins but a set of stock comedic figures. The final character, "The Modern Flapper," was juxtaposed to the rest, the program notes explain, "to bring out the sweetness and loveliness of the old pictures in comparison."[43] All the parts were played by women, so the play caricatured and subverted gender stereotypes at the same time that it critiqued the popular cultural icon of the age.[44]

The most self-conscious effort to base community unity on the heritage of the past was the "Historical Tour of Broome County" initiated by the Home Bureau. Designed to "perpetuate local historical points of interest," the tour was cosponsored by local historical societies, patriotic organizations, the Broome County Chamber of Commerce, school superintendents, and the Pomona Grange. The published guide for the tour, which took place on July 13, 1928, focused on the "early settlers" and the oldest landmarks. Although most were white men whose ancestors had arrived in south-central New York soon after its native inhabitants had been driven out during the Revolutionary War, the section on Maine added, "forget not the pioneer homemakers, for they created the school taught in a log cabin here by Betsey Ward in 1802."[45] Nanticoke Valley women used the heritage of the past to claim a public place in their community. The next year the Maine group conducted a "clinic for preschool children," an up-to-date version of social mothering that they traced back to the origin of the town.[46]

"What the Home Bureau Has Meant to Me and My Community"

In the mid-1920s, the Dairymen's League and the Broome County Farm and Home Bureau solicited essays on "what the Home Bureau has meant to me and my community."[47] The five letters from Nanticoke Valley residents that have been preserved all come from West Chenango; women who lived in East Maine joined the group in the adjacent upland hamlet rather than traveling down into the valley for meetings. Their letters lack the sweeping feminist language of some letters from Binghamton, such as Helen F. Briggs's testimonial that the Home Bureau "is the only medium through which the housewives can get in touch with progressive modern education....It helps women to find themselves...and use the talents they posses [sic]....It increases their self respect and standing in their community." On the other hand, the Nanticoke Valley letters focus on what women have done through the organization, not on what it has done for them. Florence M. (Mrs. Milo A.) Johnson cited increasing involvement in political affairs: "Women are taking an active interest in politics. Members have sent petitions to officials. More women are getting out to vote." Other letters describe the concrete improvements women made in their own lives. After enumerating the new household practices she had learned, Kathleen Stewart declared, "Before I was scarcely acquainted with my neighbors; now I know and love them well." Margaret Green (Mrs. George F.) Kolb agreed that gathering with neighbors and friends was the most important benefit of Home Bureau membership: "One returns to the duties of the ensuing days with a more cheerful attitude toward her tasks and a higher degree of efficiency....But there are some of the members to whom the Home Bureau means even more than this. To them it is a taste of recreation and association outside of their own home." By drawing women into community activities, it promoted "less strife and more social life."

Several letters emphasized that the Home Bureau served everyone. Belle North Johnson said that "the instruction and use of pressure canners" had benefited "not only the members but also those outside the bureau." Florence Johnson praised the changes that had occurred in West Chenango as a result of Home Bureau work: "The Ladies' Aid Society has been strengthened by the splendid co-operation. A fine Community House has been built." The Home Bureau held an annual fair and a community Thanksgiving dinner, sponsored recreational and cultural events, furnished aid to a burned-out family, and ensured that hot lunches were served in the district school. "Foreigners are among those having a share in the work," Florence Johnson observed. This phrase suggests that immigrants were being drawn into community-wide programs, rather than becoming the objects of charity and instruction. But Belle North Johnson expressed a more condescending attitude: the Home Bureau "served as a vehicle through which many of our foreign people have been reached and brought in and are learning how, and in better ways, to do the necessary things in home life." The assumption that immigrants were ignorant of, or deficient in, the basic skills requisite for housekeeping and childcare was common in the 1920s, and its appearance here suggests that the contact between natives and newcomers that took place in the Home Bureau was conducted on an unequal basis. Only informal visiting and work sharing would, over time, generate the familiarity required to dispel such prejudices.

Valuing Women's Labor

The Home Bureau section of the *Broome County Farm Bureau News* con-
tained articles on women's work and family responsibilities as well as reports
on local groups' activities. The Home Bureau agent who edited this section
selected from materials provided by the State Home Bureau in Albany and
the Home Economics Extension Service in Ithaca. These articles addressed
what Alice Ambler and Ann Duncan Phillips believed were farm women's
most pressing problems and proposed solutions they thought would appeal
to members. Women in the Nanticoke Valley shared some, but by no means
all, of these ideas and held firmly to their own priorities.

The information on labor-saving devices that regularly appeared in the
Home Bureau section was intended to persuade farm people of the advantages of household improvements. An article published in conjunction with
the demonstration of washing machines at the Broome County Farmers' Picnic in 1919 illustrates home economists' enthusiasm for efficiency through
modern technology: "The power washing machine represents perhaps the
greatest labor and time saving device that can be procured for the home.
Persons who are using them say they pay for themselves many times over
in a short time. What would you think of a farmer today who cradled his
wheat because he thought he could not afford a modern machine? It is just
as inefficient for the farm women to spend back-breaking hours over the
washboard when the work could be done easily by machinery."[48] The gap
between this recommendation and conditions in Broome County is glaring;
many farms lacked modern machinery because they could not afford it, and
few had electricity or the free-standing power generation systems that were
necessary to run washing machines. Even more telling, no household technology would "pay for itself" unless the time saved were devoted to other
income-producing work—or, as Elizabeth Wheaton Graves did, a woman
took in laundry to pay for her machine. Industrial measures of efficiency
were strikingly inapplicable to farm households. Farm people in the Nanticoke Valley did not share this capitalist calculus regarding investment and
its returns, much less see the logic of treating routine household chores as if
they were income-producing labor.

Home economists saw the house as left behind the farm in the march of
progress. A typical article lamented, "The women of today perform the hard
muscular work of the home very nearly in the same way that their grandmothers did."[49] Editorials took an admonitory tone, making such declarations as "It should be the rule to add some labor-saving equipment to the
house every time one is added on the farm."[50] Home economists assumed
that rural residents understood the advantages of mechanization on the
farm but had neglected to apply them to the household. Equally important,
they assumed that women were physically overburdened by housework and
that men did not appreciate the difficulty and importance of their labor. In
advocating the purchase of household equipment, home economists sought

parity between women's work in the house and men's work on the farm. Farm people did not share that gendered dichotomy in their own working lives, so they found this rationale jarring.

The articles on water systems published during summer and fall 1920, when the Farm and Home Bureau was demonstrating them across Broome County, were directed at both women and men because they served both the barn and the house.[51] According to a 1919 survey of "farm home conditions for women" made under the direction of the USDA and conducted by the Home Bureau and home economics extension agents, who analyzed and reported on the returns from New York, the majority of farm women regarded running water as the most important household convenience.[52] An advertisement printed in the *News* referred directly to the survey results: "Nine wives out of ten on the farm, when asked which of the city advantages they want most, will answer—'a modern water system.' Dishwashing, clothes-washing, scrubbing and cleaning—the hardest tasks are made easier. And think of having a modern bathroom!"[53] The standard of comparison, tellingly, is urban residences served by public utilities rather than technologies in use on the farm. Another advertisement shows a husband and wife admiring their newly installed pump and depicts him saying, "Now, Mother, we are through carrying water." The text comments, "Water, all essential in life's daily struggle for existence and success, exacts a heavy toll from every member of the family. Under old-time conditions no one escaped, and the burden has rounded the shoulders of women and children for years."[54] Here the husband's sense of shared responsibility for the task of carrying water is voiced in the first person plural, "we," but the emphasis falls on the burdens imposed on women and children. Showing the gendered assumptions implicit in this way of thinking, a Home Bureau editorial remarked, "It's the woman who has to keep up the agitation for a home water and sewerage system."[55]

Some Nanticoke Valley families participated in this campaign. Delia Whittaker Barnes urged that the installation of water systems be made a priority household management project for 1921.[56] An article in the farm section of the *News* reported that "Venley McGregor of Maine has a very practical irrigation system arranged on his farm....Besides this he uses his equipment to furnish hot and cold water for the bathroom."[57] Running water in the house is presented as an extension of the system that pumped water to the market garden and the henhouse. Water systems were central to the dairy process, which women and men shared most fully. Women who milked alongside their husbands and were responsible for washing and sterilizing dairy utensils knew how useful it was to have a supply of running water in the barn, milk room, and kitchen. Electrical power systems also served both the barn and the house. An advertisement for a Delco electric "lighting plant" captured this linkage perfectly: "It does the milking, separating, churning, grinding, washing, ironing, and sweeping."[58]

The work of women and men in the household, barnyard, field, and pasture was more integrated than distinct, but home economists ignored the actual intersection of these domains at the same time that they argued for greater parity between them. Ringing declarations such as "conveniences for the household are just as important as those on the farm" and "will pay for themselves in increased comfort and health for the whole family"[59] fell apart as soon as farm people scrutinized them closely. In "Make Every Home Mechanically Convenient," Jennie C. Jones argued, "To be sure, it is on the farm and in the barn that the income is made, but it is in the home that lives are being made or marred."[60] Jones acknowledged that the primary reason why farm families had not purchased more household conveniences was financial; farm income was limited, and investments in equipment that increased income took priority over expenditures for conveniences that offered comfort. Proclaiming the moral superiority of the values of the home over the financial imperatives of the enterprise showed the bankruptcy of this line of argument. Within the financial logic used by farm families, it was impossible for home economists to demonstrate that household conveniences would "pay for themselves." Often they resorted to a critique of the economic calculus itself, implying that farmers—whom they presumed were men—who focused entirely on costs and profits were base and selfish. One Home Bureau editorial began with the proposition that "the home, not the barn or the field, is the real center of the farm."[61] Another concluded, "A farm is a place of business. Yes, but every business is but a means to an end. That end is—home! The home is dependent on the business, but the business exists for the home."[62] Farmers like Duane Barnes, whose short speech about the sacredness of the word *home* had persuaded the men to call the organization the Farm and Home Bureau, would readily endorse that opinion. In practice, however, the farm supported the family. In Nanticoke Valley residents' eyes, the fundamental problem with home economists' formulation was that it attempted to divide the farm family in twain, rupturing the fusion of property, labor, kinship, and marriage that was at the core of their value system, reducing farming to a capitalist enterprise, and evaporating the home into an immaterial sentiment.

Home economists tried a variety of strategies to get around the stubborn fact that, by the financial calculus that controlled farm families' decisions, purchases of equipment and machinery that saved labor in the household amounted to consumption expenditures rather than productive investments. The crux of the problem was that housework had no value within the terms of the capitalist economic system.[63] New tools might enable women to perform their customary tasks with less physical effort, but their increased productivity did not yield a measurable rise in income. Home economists knew that there was something fundamentally wrong with the devaluation of women's household work, but they rarely made a radical critique of the gender inequalities inherent in capitalist economic relations.

At other times, home economists tried to demonstrate that women's household labor had measurable economic value. The only way they could do that, however, was to calculate what it would cost to purchase the things farm women made and to hire servants to perform the services women provided for their families.[64] This approach was not satisfactory because the goods and services that rural women provided were seldom available for purchase and domestic servants did not perform the whole range of tasks that farm women did. One article totaled up the concrete results of a farm woman's lifetime of labor. This farm woman

> HAS NO BANK ACCOUNT. She never "earned" any money. She lived on a...farm. She is somebody's mother. Maybe your own. She has earned nothing. No, but in her 30 working years she has served 235,425 meals; she has made 3190 loaves of bread, 5930 cakes, 7960 pies, canned 1550 quarts of fruit; she has raised 7660 chickens, and churned 5460 pounds of butter; and has put in 36,640 hours of sweeping, washing and scrubbing. At a fair price the work is worth $115,485.00. But she has no bank account to show for it. How do you express the...farm woman's contribution to her family's and the nation's wealth?[65]

Because women's household labor was not recompensed, women lacked the independent financial resources that were becoming increasingly necessary under the economic conditions of modern life.

An alternative approach to the question is suggested by a number of articles on the advantages of household conveniences and labor-saving devices that appeared in the *Broome County Farm and Home Bureau News*. These articles maintained that these investments would enable women to devote more time to working in the barn and the fields. In her annual report for 1919, for example, Alice Ambler stated, "In order that women might more efficiently perform their household tasks and thereby have more time in which to help with the work outside the home, sew, read or visit, the agent has encouraged the use of power machinery in the home and the utilization of labor- and time-saving devices." Although the Home Bureau agent described the work women did outdoors as "helping" the men, she listed income-producing work first, before gender-marked work such as sewing and leisure-time pursuits such as reading and visiting.

This list reflects the priority that farm women gave to their moneymaking activities. Articles on the advantages of the fireless cooker were especially explicit. One was headlined: "Cook Helped Out of Doors All Day But Had Hot Dinner."[66] Song lyrics by Mrs. Michael Donahue of Cascade Valley, an upland area near the Pennsylvania line east of Binghamton, included this verse:

> *I have a fireless cooker, too, by which I set great store*
> *I start my dinner in it—go out and do the chores,*
> *Then hustle around like sixty in the garden, field and barn*
> *Since I have a fireless cooker.*[67]

An editorial titled "Do It by Machinery" that appeared in the Home Bureau section in March 1921 stated the case in more general terms: "Just at this season when the women are doing so much work outside, it seems that everyone should realize more strongly the necessity of machinery in the home.... Life on the farm is a partnership, and whatever adds to the productive power of either of the partners, increases the earnings of the farm. As a business proposition, the farmer must save the strength, and increase the efficiency of his partner just as he does his own."[68] The crux of this argument is women's participation in income-producing farm labor. This editorial assumes not only that men evaluate any proposed innovation "as a business proposition" but also that they recognize their wives as working partners in the enterprise. Significantly, all the articles that advance this argument originated at the local level. Farm women and the agent who worked directly with them approached this question differently than home economics professionals did. Firmly rejecting the gendered separation between farm and household, they posited an alternative model of the family farm as a partnership in which women and men shared productive labor.

The question of who had the power to allocate farm family resources and in whose interests those decisions were made was implicit throughout these discussions. Home economists assumed that men made the final decisions about whether to purchase household conveniences, just as they did about investments in farm machinery. Home economists' analogies between farm and household technologies and their use of a capitalist economic calculus to evaluate productivity gains were, in part, attempts to appeal to men. On the other hand, they sought to change the balance of power in farm family decision making, increasing women's say so that their interests would be better served. Often they adopted a gendered model of decision making that corresponded with their gender-segregated model of labor; the wife should make decisions about the home, although the husband makes decisions about the farm. This model was inapplicable to most farm families, and women in the Nanticoke Valley upheld an ethos of cooperation that was the obverse of home economists' view. Women who regarded the entire farm as their domain willingly put the barn before the house and had little hesitation about speaking up in family councils.

Many home economists attempted to advance women's interests by increasing the resources over which women could exercise independent control because they were produced through women's own enterprises and yielded a separate income. In this endeavor, they were responding to farm women's constant, often desperate pleas for help in finding ways to turn their labor into cash. In the 1919 USDA survey of "farm home conditions for women," women overwhelmingly asserted that what they needed most was an increase in farm income in general and new ways of obtaining an economic return from their own labor in particular.[69] This strategy was modeled on poultry operations, which had traditionally been under women's control. It was workable in some circumstances, especially when rural residents had direct access to urban markets. Maple syrup, jam and preserves, high-quality butter, buttermilk, and cottage cheese were all promoted as ways women could earn money of their own. Serving as a kind of clearing-house for farm women, home economists publicized particularly successful enterprises.[70]

This strategy was not readily applicable to the large-scale farm operations in which women worked jointly with their husbands. Indeed, agriculture experts assumed that commercial poultry operations were conducted by men. Articles on poultry raising referred to farmers in the male gender, even though photographs of poultry demonstrations show mixed audiences. The American Farm Bureau Federation proposal for the cooperative marketing of eggs, in contrast, acknowledged women's customary role in the poultry operation. The plan provided that "The woman...be given a dominant part, because of our recognition of the fact that the marketing of eggs has heretofore been left, in a majority of cases, to the control of the women members of farm families, and because the proceeds from the sale of eggs have been used in direct home expenditures by such women members and are directly concerned with the comforts and standard of living in such homes."[71] This provision of the plan reflects the influence of Home Bureau leaders, who were concerned that a masculine takeover of poultry marketing would result in a shift of resources and power away from women.

Local Home Bureau groups chose to learn skills that would help them manage money and make financial decisions, either independently or in cooperation with their husbands. The programs on "Business Methods for Women" presented at Homemakers' Conferences by the Cornell Extension staff raised the question of power quite directly. The report on a conference held during Farmers' Week in Ithaca asks bluntly, "If you have a joint income, do you have a joint bank account? If joint labor has earned the property, is there a joint deed to the property?"[72] Nanticoke Valley women shared these concerns because they regarded women as farm partners and thought that financial arrangements should recognize all family members' contributions to the work as well as facilitating the continuation of the enterprise.

Items published in the *Broome County Farm and Home Bureau News* presume that women participated in farm family decision making and financial planning on a day-to-day basis. Most articles recommend that the farm and household be conducted as a financial partnership.[73] Some hint at tensions between women and men over small personal expenditures: men may worry that women spend too much on clothing, and women feel that men spend too much on tobacco and alcohol.[74] But even these articles assume that each spouse has access to the couple's joint funds and is aware of the other's purchases. Home economists' nightmare—that farm women were drudges who toiled indoors and out but had no power over decision making and were even denied access to cash—finds little support in the evidence from Broome County. Nanticoke Valley women shared home economists' advocacy of joint decision making, but they premised their right to have a say in financial management on their active participation in farm labor.

Nanticoke Valley women did not respond positively to Home Bureau exhortations for them to become consumers, albeit well-informed ones. They regarded such advice as inappropriate for rural women and even for the wage-earning women they knew, who worked in the factory to support their families. They listened more sympathetically to Martha Van Rensselaer, who emphasized the role of women as the money-savers in the household when she spoke on the "Economic Value of the Home" at the countywide Home Bureau rally held in Binghamton in June 1920.[75] Her address assumed that farm women's sense of the economic value of their labor was rooted in their household production for use as well as in their participation in income-producing labor. Whether they were adopting household conveniences and labor-saving devices to spend more time in the barn and fields or keeping household accounts to reduce consumption expenditures and save money for investments in livestock and farm machinery, farm women had a firm sense of the value of their work. They based their participation in decision making on their flexible but substantial contributions to the necessary subsistence and income-producing labor of the family farm.

Conclusion

Gender, Mutuality,
and Community in Retrospect

From the onset of the Great Depression through World War II, most people in the Nanticoke Valley held on to the land they had inherited or managed to purchase. Whether their ancestors had come before the Civil War or after World War I did not matter much when everyone was struggling to grow enough food and cut enough wood to stay warm and well nourished. Workers from European peasant backgrounds had bought farmland during the 1920s, in part because employment in the mines and factories was unstable even in relatively good times and in part because they did not want to be entirely dependent upon employers and landlords. When unemployment soared during the 1930s, they could work for themselves on the land they owned and house their families without having to pay rent. Those longtime residents who had kept their farms while sending family members to work in the city also devoted more of their efforts to subsistence production and small-scale, market-oriented agriculture. Some people from the Nanticoke Valley who had moved to the city returned to live with relatives rather than face destitution or depend on the dole. Adult children brought their own children back to their parents' farms; divorced, widowed, and deserted individuals of both sexes relied on the security that farming offered. Diversified farms demonstrated their value, providing a hard-earned modicum of security, but specialized commercial farms could no longer expand at the same rate as before even though cheap labor was more readily available. The trend toward stratification between these two classes of farming families was halted, although not reversed, during the protracted period when holding on was the best that most rural residents could manage.

Getting By during the Great Depression

People in the countryside coped with their common predicament through customs of cooperation, sharing work and tools in the annual round of farm labor and relying on mutual aid to ensure that most families had the bare necessities. In his second inaugural address in January 1937, President Franklin D. Roosevelt acknowledged that "I see one-third of a nation ill-housed, ill-clad, and ill-nourished."[1] Those who lived through the Great Depression in the Nanticoke Valley did not count themselves among these impoverished Americans. They concurred that they had suffered less than urban dwellers, in part because they were accustomed to getting by with little and in part because they were never hungry or cold. Even though some families had a bit more than most people and others lost their farms to foreclosure, everyone seemed to be in a similar position. Class distinctions paled when the prices farm products brought fell below the cost of producing them and nonfarming and farming families alike relied on big gardens. The revitalization programs that rural residents had adopted during the 1920s, especially the building of community halls for recreational activities and the federation of Protestant churches, stood them in good stead when disposable income disappeared. Strong local institutions were based on and reinforced long-standing practices of mutual aid. People shared produce and labor outside the cash nexus both within and beyond extended kinship networks. The mutual aid that no one would call "charity" when its recipients were neighbors rather than strangers proved its value in what became a chronic emergency. In the 1930s, the public schools provided hot lunches and medical checkups to all children, and voluntary organizations such as the Farm and Home Bureau helped set up local agencies to administer state and federally funded welfare programs.

The Depression reinforced rural residents' long-standing economic and social practices, as well as underlining the validity of their critique of a capitalist market in which processors and distributors had more power than producers. Farmers acted collectively through the Dairymen's League to secure prices for their milk that, at a minimum, repaid the cost of producing it. The stability that was achieved in the early 1920s by the introduction of the pooling plan and the establishment of cooperative processing plants did not prevent rapid corporate consolidation in the industry and downward pressure on milk prices. Even before the 1929 financial crisis, Congress was considering legislation to make commodity prices more equitable. As the farm crisis deepened, R. D. Cooper's history of the Dairymen's League notes, "the increased interest of women in the organization was evidenced by the presence of a thousand farm women" at the 1929 annual meeting in Syracuse.[2] That year the league formed a Home and Health Education Department under the leadership of Vera McCrea, and its session at the 1930 annual meeting in Albany was attended by "1000 League women and 500 men."[3]

This grassroots, gender-integrated mobilization indicates the rising level of concern about the consequences of the economic crisis for farmers. In fall 1932, the urban market was flooded by non–Dairymen's League milk, and prices fell precipitously. By 1933, it became clear that only government intervention would bring price stability, and New York State set up a Milk Control Board to regulate wholesale and retail prices. Nanticoke Valley farmers participated in the debate. In February 1933, Ralph Young noted in his diary that he, his father, and his brother went to "a Dairymen's meeting" in Union Center. "Paul Smith spoke on the plan for State control of the milk industry. Milk prices are too low."

That August, some farmers in upstate New York called a strike against the price schedule imposed for different classes of milk by the Milk Control Board.[4] This direct action was taken without the agreement of the Dairymen's League, making it a wildcat strike.[5] Farmers who stopped sending their milk interfered with those who chose to honor the contract, trying to enforce a unanimity that did not exist. Indeed, the conflict pitted farmers against one another rather than against the milk dealers. The Youngs, along with others from their neighborhood, decided not to participate because they feared that taking action when farmers were not united would result in chaos rather than higher prices. Anticipating trouble, they made a new arrangement for selling their milk. Ralph Young recorded on August 1, "We began delivering milk to Nathan Young at Union at night instead of to Cloverdale Co. of Binghamton where it has gone for several years. Special permit from D. League." This retail dairy (whose proprietor was no relation) delivered pasteurized milk to local customers each morning. The farmers had permission from the Dairymen's League to make this change and still paid a commission to the cooperative. They were beset by difficulties five days later: "Milk strike is on. Men from outside came and stopped all milk deliveries from Union Center. A crowd gathered to stop our milk at night but with the aid of 7 Endicott policemen we went through okay." Ralph identified those who tried to stop the shipment as outsiders because other farmers in their neighborhood also continued shipping milk. When we discussed this strike in an interview, however, Margaret Young Coles reminded her father that their own milk had been dumped near Scherter Road by people whom they knew. "The rest of the farmers thought one way, we thought the other way," Ralph admitted, acknowledging that they were in the minority. The state intervened on the side of those who sought to comply with their contracts. The next day, according to Ralph's diary, "Milk went through from Union Center by the...truck going to the farms and getting it with guards along. We delivered milk under the protection of policemen tonight. Regular Grange meeting was to have been held at night instead of tomorrow night but not enough there to hold it." The Grange, too, was torn asunder. A week later, "Milk strike dying out. We went with our milk alone tonight, the first time this week that the police were not here."

The divisions that marked this strike represented a serious breakdown in farmers' solidarity. Farmers like the Youngs, who were selling fluid milk directly to nearby retail dairy companies, did not participate. Because the Dairymen's League had not called the strike, those who sold through the co-operative were also divided. In the northern end of the Nanticoke Valley near Whitney Point, the Northfields continued taking their milk to the Dairymen's League plant while other farmers went on strike. Carrie Northfield explained, "There were two groups, those that were for the strike and those that were against. We didn't go along with the strike, because we didn't think we could afford to throw our milk down the drain, but some of the people down the road" who were on strike tried to dump their milk. When her older brother "went to the milk factory, somebody hit him with a plank, even, and it got to be quite violent." Her younger brother Richard explained the politics involved; their father "was one of the local officers of the Dairymen's League, so he couldn't very well strike, even though he wanted more for his milk." Carrie concluded that the split among the neighbors "all blew over" afterward, although "it was very vicious" at the time. Similar conflicts occurred near Maine village. Sarah Briggs Brookdale recalled the bitterness that had arisen between her family and their next-door neighbors during the strike: "I never quite knew just what happened, but there were hard words between the Briggses and the Watsons, because I guess my father wanted to ship milk and the Watsons didn't, and then neither of the Watson men came to do chores and Hazel didn't do the laundry, which they always did—even though, heaven knows, they needed the money. Anyway, no milk went from here. Afterwards, we always drove round the long way to get out to the main road so as not to go past their house. I don't know why that went on for so long." Hazel, who was eleven at the time of the strike, told me she thought it was "silly—stupid, really" for neighbors to shun one another because of a difference of opinion over an issue like this. Rural people "just had to rely on each other," even though they might disagree about politics from time to time.

Although the Youngs continued to be active in the Grange and the Grange League Federation and attended Dairymen's League meetings in Binghamton through the next winter, they left the organization temporarily in 1934 when the retail dairy in Endicott was no longer covered by the league contract. In our interview, however, Ralph emphasized the family's allegiance to collective action. "When things had settled down" they rejoined the Dairymen's League and sold their milk to Dairylea, the cooperative's processing company. "We lost money" by selling milk through the league, "but we were willing to sacrifice for the principle of doing what we thought was best." "The principle was to cooperate with the neighbors," Margaret Young Coles clarified. "We didn't want to have any trouble with the neighbors," Ralph concurred. "We were willing to overlook quite a lot"—even the fact that others had dumped their milk. "We had to sacrifice a little; we kept on wanting to get back in, and got back in and everything was smooth."[6]

Notwithstanding this division among farmers, it seemed to some Nanticoke Valley residents that the nation as a whole returned to rural values during the Depression. After Congress passed the Agricultural Adjustment Act to regulate farm prices, the issues that were not resolved by the milk strike were addressed by joint state and federal action. Dyed-in-the-wool Republicans from the party of Lincoln saw Franklin Delano Roosevelt as continuing that democratic tradition (with a small *d*), so they voted for Democratic candidates in statewide and national elections without becoming affiliated with the Democratic Party that was centered in urban Broome County. Town and county elections turned on local issues, especially school consolidation and road construction. Men such as George W. Young and Edward F. Vincent traded positions on town boards and the Broome County Board of Supervisors. In Maine and Nanticoke, women—some of whom were married with grown children, but several of whom were divorced and raising children by themselves—served as town clerk, tax assessor's clerk, and county court clerk, doing the administration and dealing with disgruntled residents.

Home production, cash-saving, and making do were key survival strategies during the Great Depression. The Home Bureau emphasized activities that had long served farm families and rural communities.[7] In response to cuts in its annual subsidy, the organization trimmed its budget, stressed community service, and relied even more heavily on volunteers. Its focus on home production stood women in good stead. Indeed, such topics as how to reuse flour sacks had been popular well before 1929.[8] The agricultural depression had begun after the First World War; what was different about the 1930s was that the predicament of farm families became publicly recognized and was shared with city dwellers. The projects that members had favored, which focused on making do, now enjoyed greater support from home economics professionals. Programs on housework emphasized saving money and materials rather than just saving time, as such titles as "low-cost meals" and "reconditioning furniture" indicate.

In the 1930s, Home Bureau leaders reinforced the ethic of community service. They reminded local groups to be inclusive, focusing on rural residents' shared needs and values rather than on their differing social status or circumstances. Delegates to the 1934 annual meeting were cautioned, "Use care in meetings that they do not become 'clubby.' Extend the work to non-members and serve as many non-members as possible."[9] Sending baskets filled with food and other necessities to people who were "too proud to ask for help" was an important part of Home Bureau work. The Glen Aubrey, Maine, and Union Center groups all continued to organize a variety of free or low-cost educational, social, and recreational activities.[10]

At the same time, however, women from activist families continued to focus more on the farm side of Farm and Home Bureau programs. Maude Young rarely attended local Home Bureau meetings and never served in a leadership position, but she joined Ralph at many Farm Bureau events. Indeed, in 1934 the couple celebrated their anniversary by going on a Farm Bureau tour of Broome County. The Youngs and Ketchums were especially interested in "poultry schools," which combined information with demonstrations, and in July 1939 "Louis, Norma and Merton Ketchum, Father, Maude and I went to a poultry tour" that stopped at the McGregors'.

The Grange continued to attract married couples and involve entire families. The Juvenile Grange, which the Union Center group formed in 1932, was (like the later 4-H) designed to make farming appeal to children, but in practice it tended to draw mothers with young children into Grange leadership. As the Youngs' children matured, Maude accompanied Ralph to an increasing number and variety of Grange functions. By 1934, she was regularly attending meetings of the Pomona Grange across the county. She accompanied Ralph to state conventions every year, and in 1942, she served as a delegate when the elected delegate was unable to attend. The contrast between the Grange and the Dairymen's League, whose leadership remained exclusively male despite the active participation of women, became especially visible during these years. Maude accompanied Ralph only to social events held by the Dairymen's League, such as a dinner "held for officers and their wives" in Endicott in 1939.

The Community Day picnics held in Union Center continued through the 1930s. The afternoon featured a baseball game between the married men and the single men, complemented in 1935 by a "softball game between the married women and girls." In 1933, the event was called a Community Picnic and Old Home Day, and those who had moved away joined those who had stayed behind; about two hundred people attended. Old Home Day was to the community what a reunion was to the family, an attempt to transmute into a symbolic but enduring form what had once been an immediate and intensive association but was no longer because people who had grown up in the community had scattered throughout the urban areas of the county.

Postwar Transformations

The depletion of resources that surviving this protracted crisis entailed had serious long-term consequences. Most became visible only after World War II, which brought prosperity back to the rural community as factory employment soared, prices for farm commodities rose, and the national economy revived. Wartime demand for boots manufactured by Endicott Johnson workers; uniforms sewn in Binghamton; and armaments that used components engineered and fabricated at IBM, General Electric, and Singer-Link was complemented by federal control of the commanding heights of the economy, including production, labor, and finance, which helped to protect the interests of organized workers and farmers as well as ensuring corporate profits. In agriculture, the policy of price supports and marketing boards that had been adopted during the late 1930s was utilized to promote dairy production at prices that made it possible for some farmers to expand their operations. To those who did not remember the economic downturn that had followed the First World War, especially the precipitous decline in farm commodity prices and the rapid inflation in the prices of farm inputs as well as essential consumer goods, it seemed that the wartime boom would usher in sustained prosperity for all once peace came.

The outcome of two difficult decades and five boom years, however, was quite mixed. Economic inequality increased markedly, and many farmers who had hung on but had been unable to accumulate capital found it impossible to meet the requirements of commercial dairying. The consolidation of milk processing accelerated during the war, and now milk tanker trucks came to farms once daily—provided that they were located on paved roads and produced enough milk to make a stop worthwhile. Indeed, isolated farms that produced large quantities of milk might find themselves passed by because their neighbors were too few and far between to justify a route. Gatherings of farmers at the creameries or milk pickup stations vanished overnight. Still more serious were the escalating requirements for dairy farms that sold fluid milk. To pass the inspections that contracts specified, barns had to have concrete floors, milking machines, and separate milk houses, which guaranteed cleanliness and facilitated pickup. Eventually they had to have bulk tanks and pipeline milking systems. Year-round production became essential, so farmers had to buy winter feed and grow more corn for silage. The capital necessary simply to stay in business was far beyond the reach of farms that had barely survived the 1930s. Land values and average dairy receipts were not high enough for most farmers to borrow money, although federal programs enabled those who already operated on a fairly large scale to expand and purchase new equipment.[11]

The sudden disappearance of the market for farm-made butter, buttermilk, and cottage cheese; eggs and chickens; and fresh vegetables and fruits proved equally fatal to those farmers who had depended on small-scale, diversified operations and sold their produce in nearby cities. The public markets in Johnson City and Endicott closed after the war, replaced by chain supermarkets that sold fresh, frozen, and canned vegetables and fruit from California and Florida and even from elsewhere in New York state. Potatoes from Idaho, Maine, and Prince Edward Island in Canada replaced those grown in Broome County. The poultry industry consolidated rapidly. Supermarkets contracted only with large-scale producers, who could guarantee regular supplies that would meet the demand of each store. The small grocery stores where Nanticoke Valley farmers had once taken their butter and eggs succumbed to the competition.[12] The demise of marketing outlets for locally grown food seemed the price of progress. Indeed, farm families found that they could buy food more cheaply than they could grow it for themselves. Those who could afford to care about quality and buy freezers continued to tend their gardens and orchards to furnish their own table, but those who had to worry about using their time and money most efficiently concentrated on what was profitable and made weekly trips to the grocery store, just as urban dwellers did.

Few of the families who combined farming with wage-earning were able to stay in agriculture after the end of the war. Farmers who started out during the 1930s were less likely to succeed than those who had started out earlier and benefited from the prosperity brought by World War I. Immigrant families, in particular, had a hard time. Their fledgling enterprises could support one family, but not two, and setting up a partnership required more capital than most had and more credit than they could obtain.

Farm succession was as serious a problem in the aftermath of World War II as it had been after World War I.[13] Sons and daughters who had joined the military or taken jobs in war plants traveled across the country and around the world, marrying people they would never have met otherwise and deciding to settle in faraway places they had visited. The geographical and cultural distance between this generation of city dwellers and their parents on the farm was much further than that between the Nanticoke Valley and Endicott, Johnson City, and Binghamton. Married children brought their spouses and the grandchildren to visit but seldom stayed. Only a few youths returned and took on the task of continuing or rebuilding a farm. More shared a house with and cared for their aging parents after the cows and machinery had been sold and commuted to work in the Triple Cities. Many of the farms that operated at a large enough scale to make the postwar transition successfully had no heirs who wanted to carry on. That fact did not particularly disturb couples who had gone into farming for themselves, with little or no aid from their parents and kin, during the late 1920s and early 1930s. Their understanding of the enterprise differed from that of farmers who had followed in their parents' footsteps and inherited going operations. The generation that came of age during and after World War I had enjoyed a taste of personal independence and, to a remarkable degree, raised their children to make their own decisions. They did not expect them to carry on. The notion of the family farm as a multigenerational enterprise had already been undermined, and by 1950 more families were able to look back at a century of farming on the home place than forward to a future in agriculture.

The collapse of diversified, small-scale farming, especially on the hills away from the main roads, had subtle but serious consequences for the larger, commercially oriented farmers as well. The time-honored practice of changing work disappeared when farms were further apart and fewer neighbors had available labor. There had always been a delicate balance between farmers who could afford machinery but needed extra hands at haying and harvest and those who depended on others' machinery but compensated by sending extra workers to the crews that moved from one farm to the next in an open-country neighborhood. During World War II, harvesting and threshing became a commercial enterprise; those who owned machines hired wage laborers for the season and worked for farmers throughout a larger area on contracts for so much per acre. Larger machines that per-formed more than one task, such as combines, lessened the need for extra hands to do the manual labor required to keep up. Hay rakes, tedders, and balers and harvesters that cut and chopped corn for ensilage mechanized tasks that had previously demanded many hands to follow after a horse-drawn or tractor-pulled machine. As the scale of farm operations increased, too, the urgency and magnitude of the harvest intensified, and farmers felt compelled to buy their own machinery so they did not have to wait their turn and worry about the risks of rain or frost damage to their crops. These trends were mutually reinforcing. As larger farmers bought machinery, they withdrew from neighborhood arrangements for changing work; as neigh-bors who continued to cooperate in these tasks became fewer and farther between, more farmers had to buy harvesting machinery.[14]

As Edmund Smith lamented, "neighboring just wasn't what it used to be." The dissolution of open-country neighborhoods accelerated as families moved away and the practice of cooperating in farm work declined. First, "as the farms decreased there just weren't the people" to change work in haying and harvest.

> In the second place, one of the big problems was that everybody wanted to fill silo the same day. I don't know, maybe they became affluent enough that everybody got a tractor and they could get a blower. At one time, a whole bunch of 'em got together and bought a blower and a corn cutter or harvester, but that soon broke up, because it always left somebody having to begin real early and somebody was left to finish after it was real late. So pretty soon somebody stepped out and bought themselves a second-hand blower and so forth, so it soon got down to pretty much...Well, they still generally got some help to haul the corn up with wagons so they could blow all day long, but even that began to peter out after a while, and everybody began to get their own equipment.

Then it was every family for themselves.

In turn, the demise of neighborhood work-sharing undermined the orga-nizations and institutions that had sustained farm families. Women remained integral to dairy production, but as commodity associations and marketing cooperatives became more bureaucratic, larger in scale, and enmeshed with the state, they became more male-dominated. The proliferation of federal and state agencies dealing with agriculture was part of this process. Farm-ers such as the McGregors and the Youngs became involved with New Deal programs, but did so as representatives of farmers like themselves rather than as representatives of their locality. The rural electrification program, like the cooperative that had strung power lines up the Nanticoke Creek Road a decade and a half earlier, involved entire neighborhoods. But other groups were organized only at the county and state levels. For example, Ralph Young was on the Soil Conservation committee, and the Young farm participated in a government-sponsored soil conservation program in 1937–1938. Ralph attended the first meeting of the Broome County Agricultural Planning Committee in 1937. In 1939, when the USDA distributed grass seed to farmers after a severe drought, he made a tour of the county to inspect the reseeded meadows and pastures. By 1940, he and Stacey belonged to the Fivetown Milk Producers' Association, and Ralph served on its board of directors. Ralph also joined the Greater Endicott Post-War Committee on Agriculture. These activities led him beyond the Nanticoke Valley and, eventually, out of Broome County entirely when he went to work for the federal Production Credit Association. As farm organizations came to represent only the specialized, better capitalized farms, the class stratification that had been held at bay in the Nanticoke Valley finally fractured local society. The fortunate few farming families who stayed in business became more connected with their counterparts elsewhere, while the rest earned a living however they could. The chronic deprivation and isolation that can still be glimpsed from the back roads was severe enough that in 1965 Broome County was included in the Appalachian Regional Commission, a federal antipoverty program.[15]

Strategies of Mutuality

People in the Nanticoke Valley held to an alternative, rural way of life, which they thought of as different from the dominant urban culture. Their daily lives exemplified the core values they espoused: gender integration and mutuality, reciprocity and mutual aid, and social equality and collective action. By World War I, rural residents had become aware that their way of life—which they thought of as ordinary, and which their parents and grandparents had thought of as quintessentially, if not universally, American—was no longer shared by most people in the country. They sought to defend it from an economic system they thought exploited farmers, a society they saw as increasingly divided by class and ethnicity, and a polity they saw as controlled by large corporations that acted to promote their special interests against those of the common people.

Those Nanticoke Valley residents who were active in farmers' cooperatives and political movements subscribed to what historians have called a producer ethic, the idea that the people who perform the world's work by laboring on the land and making useful goods are morally superior to "parasites" who shuffle papers in offices, exploit the laborers they employ, speculate in land or deal in money, and take advantage of others in the capitalist market.[16] The ideas that farmers continued to hold in the early twentieth century had much in common with Populism, which had enjoyed broad support in rural America during the late nineteenth century, and with the economic ideology that had informed the Farmers' Alliances, which welcomed women, African Americans, and immigrants along with native-born white men into the ranks of producers. In political terms, they adhered to the kind of republicanism that had inspired the party of Lincoln to uphold the principle of "free soil, free labor, and free men" at such high cost in local lives.[17] The economic and political subordination of one class to another, especially of producers to parasites, struck them as akin to enslavement by a modern-day aristocracy. Through the economic depressions and social upheavals of the late nineteenth century, Nanticoke Valley farmers had quietly, but firmly, maintained their producers' cooperatives and continued to practice the kind of egalitarianism they thought expressed basic American values.

Practices and norms of gender flexibility and integration were fundamental to this alternative way of life. In the vernacular language used by the women with whom I discussed these issues in depth and by the men I interviewed, "the family farm" and "the farming family" were interchangeable terms because farm and family were equal and inseparable. What another rural historian has amusingly, albeit accidentally, dubbed the "farmily"[18] was centered on work sharing across lines of sex and age and grounded in a common commitment to the farm as the foundation of the family. The idea that relationships between women and men could or should be defined by

Figure 12. Woman standing by a dump rake, loading hay up onto a wagon where a man distributes it. Maine, New York. Undated photograph, photographer unknown. *(Courtesy of the Nanticoke Valley Historical Society.)*

such fixed and limiting roles as "breadwinner" and "homemaker" and that school-age children ought to be "dependents" seemed alien to them. Few farm women voiced a feminist critique of the gender system that prevailed in the dominant society except as they looked back on the changes that had taken place recently when I interviewed them in the 1980s. Instead, they simply stated their alternative view as if it were commonsensical and regarded other ideas as strange and irrelevant.

Nanticoke Valley residents knew that some middle-class city dwellers thought that women and men had by nature different capacities and distinct responsibilities, but to them this conception was absurd when applied to anything beyond bearing and breast-feeding babies. No one in a farming family had much use for such urban ideas, let alone the practices that they implied. The images of women that circulated in the mass-market magazines they saw from the 1920s on came in for particular criticism. Some women told me that no self-respecting rural woman would have thought for a moment that she should resemble the lovely but conspicuously idle urban women depicted in popular advertising or the romantic stories printed in periodicals. Some men said that in their youth they did not mind women having short hair and wearing overalls, which were practical, but they would never have considered marrying a woman who looked fragile and expected to wear silk stockings; unpretentious women who were capable of doing their share of the work made the best farmers.

Everyone in the family was of necessity a contributor, and what "the farm needed" was more important than who performed which tasks. Individuals in farm families might well have different ideas about what was best for the farm and try to persuade others that the course they preferred was the right way to go about things. Individuals might bargain over the particular tasks they were expected to perform and debate how much labor and capital should be devoted to which operations. Husbands and wives did not necessarily have the same interests, just as parents and children might well have differing perspectives. But there was no prevailing pattern of work that set women apart from men. The household toil that many farm women regarded as the universal bane of womankind was less important to them than the income-producing and subsistence-oriented work they shared with men or performed on their own. The gender division of labor on Nanticoke Valley farms and in families that combined farming with wage-earning was remarkably varied. Women put the barn before the house in terms of expenditures largely because they did not regard the world as divided into *her* house and *his* barn and fields. Instead, women left the house to go to the barn and out to the fields whenever they were needed. Dairying required twice-daily, year-round family labor; poultry raising depended on women and children; haying and harvest drew all available hands into the meadows and fields; and threshing and silo-filling took over the farmyard and kitchen. These crucial daily and seasonal tasks could not be completed without women's active participation except by large-scale farmers with money to spend—or, as most others saw it, to waste—on hired help. When rural residents sent family members to work in town, the woman might run the farm while the man worked for wages, or the woman might hold an off-farm job while the man devoted his energies to agriculture. Husbands and wives might also change places, depending on their employment opportunities and the ages of their children, and shift back and forth between farm and off-farm work over time.

What is most remarkable about the history of agriculture in the Nanticoke Valley is the relatively long period during which the local preference for husband and wife to work side by side in the barn and hayfields prevailed. Indeed, this predilection was so widespread that it influenced the development of dairying in the region. Throughout the interwar period, the majority of farmers milked mainly during the summer, while the cows grazed on fresh pasture, and fed hay during the winter, when the cows were dry. In other regions of New York state, the trend from World War I on was toward intensive dairying: milking cows throughout the winter, when milk prices were higher because the supply was scarce, and drying them off during the summer, when prices were depressed by oversupply. That required feeding large quantities of silage made from corn grown on the farm and high-protein concentrates that had to be purchased. The experts favored winter dairying because it utilized hired labor efficiently; the fact that it required larger herds and more operating capital seemed to most of them signs of progress.

When they scrutinized conditions on Broome County farms, however, agricultural economists discovered to their chagrin that winter dairying was not so profitable as summer dairying, which took advantage of the meadows and pastures that abounded on upland farms and got around these farms' lack of fertile fields. Indeed, capital-intensive dairying was seldom profitable because farmers had to buy expensive feed grown in the Midwest and pay a hired hand year-round. Summer dairying relied almost entirely on family labor. Although agricultural economists acknowledged the value of the work done by unpaid family members, whom they described as the wives and children of the adult men whom they regarded as the farm owner-operators, their data demonstrate even more clearly the crucial contribution that women and children made to sustaining these enterprises. The relatively small-scale farms that predominated in the Nanticoke Valley survived and thrived because of women's active participation in milking, doing chores, and haying. When marketing conditions shifted after World War II to make larger herds, year-round milking, and the purchase of commercial feed mandatory, many local farmers were unable to continue because family labor—which most often meant *women's* labor—could no longer be substituted for capital.

Building the Base for Farm Organizations

The integral participation of women in organizing political-economic movements was equally important in sustaining family farming.[19] In the Nanticoke Valley, women were central to the formation of cooperative groups that took collective action against the milk dealers. Historians who have examined the Farm Bureau and the Dairymen's League as formal institutions have emphasized that they represented large-scale farms and were male-dominated.[20] At the grassroots, however, women held more responsible positions and did much of the work that kept these organizations alive. Indeed, when we examine the activities in which people participated, we see a balanced combination of parallel structures, in which women and men had distinct responsibilities and carried out their responsibilities in an autonomous manner, and integrative mechanisms whereby women and men coordinated their activities in mutually supportive ways.[21] The activities that sustained the Grange, the Farm and Home Bureau, and the Dairymen's League were gender-integrated. Much of the credit for that belongs to the women, who planned and conducted events in such a way that no one— least of all themselves—was left out.

Advocacy of women's full citizenship was a long-standing plank in the Grange platform. Even before suffrage, any notion that adult male heads of household represented their wives and children in politics and formal organizations was contradicted by the presence of women and young people at meetings where pressing issues were discussed and by women's predominance in church and school district affairs. Rural organizations had inclusive, democratic, and egalitarian or parallel structures of governance that placed a premium on cooperative activity by women and men. Through the sociability that they sustained and the mutual aid that they practiced, women collectively exemplified and inculcated habits of affiliation and norms of reciprocity.

Early-twentieth-century urban dwellers, whose social worlds were more sex-segregated, would have been surprised by the relatively free association between women and men in the Nanticoke Valley. To a degree that is difficult for contemporary white Americans to understand but may be familiar to members of African American communities, women were *the* community organizers, creating the webs of informal association that served as the foundation for farm organizations. They did not merely provide the food, drink, and entertainment that drew people to social events, as conventional notions about ladies' auxiliaries assume. Rather, women's inclusive practices ensured broad-based participation. Women who participated in these groups along with their husbands and kinsmen were neither defined entirely through their marital relationships nor treated simply as members of the female sex. The density and flexibility of relationships gave them room to maneuver. Alice and Maude Young not only built on their kinship ties to facilitate their participation in community life but also utilized those organizations to create modes of sociability they shared with their husbands.

In the Nanticoke Valley, kinship and neighboring overlapped and blended in ways that promoted the inclusion of immigrants as well as the relative equality of women. The mutual aid that kinship guaranteed spilled over into relations between neighbors who were not kin.[22] When immigrants bought vacant farms, they often followed lines of kinship and friendship as well. At first, the newcomers associated mainly with their relatives nearby and co-ethnics elsewhere, but open-country neighborhoods brought people together. Ralph Young and Edmund Smith both said that some native-born farmers were "prejudiced against immigrants, especially Catholics," but as neighbors worked together they "got to know one another." Personal acquaintance and common interests replaced the prejudice born of "ignorance," they concluded.

These values of inclusiveness and democracy ensured wide participation in collective action. Mutuality was a core value, rooted in the relationships between husbands and wives and parents and children within households, that extended between households and farms as well. Gender integration and economic cooperation were mutually reinforcing. As Karen V. Hansen, a historical sociologist, puts it, the key to rural women's exercise of power lay in the assertion of "we."[23] Theirs was not an individualist claim (i.e., for "me") but one that subsumed the self in the family and, for many, the farm. The power of "we" that women espoused was demonstrated by the acknowledged value of their labor, whose flexibility made it crucial to many farm operations, and advanced by the affiliations with others that women constructed through their social activities on behalf of the common good. Mutuality was both a gendered strategy for ensuring that women were not subordinated or marginalized and a social norm that women exemplified in the community. As isolated individuals, women had less power than men; above all, they seldom owned land or other productive property solely in their own names. Those who founded farms with their husbands might hold title jointly, especially if they had contributed to the down payment from their own earnings. Theirs was, above all, a partnership. Rural women recognized that reciprocity was essential to their own sustenance, as well as to others. The risk inherent in this strategy was that women might be self-sacrificing while men were selfish. But mutuality extended from the household into the neighborhood and kinship network. And, in what Mary Neth aptly called "building the base,"[24] women nurtured the social connections that supported community cooperation and farmers' organizing.

Women's Perspectives on Historical Change

When I interviewed older women in the early 1980s, they often compared their own lives to those of their daughters and daughters-in-law and put the historical changes that had occurred during the twentieth century into perspective. A few of these discussions involved two or three women, but most took place with individuals in private. What I found most striking was the degree to which these women concurred in their assessments of recent changes in women's social position. Whether they had farmed or worked in the factory and whether they were the children of natives or newcomers did not make much difference; they articulated a common view that may well have been forged in the frequent conversations about their children and grandchildren that marked their informal interactions. My questions probably prompted some to make broader statements than they might otherwise have done, but they were confident that the viewpoint they expressed was widely shared.

As these women looked at the present in comparison to their own lives, they felt that women had become more isolated, their families more private, their problems more personal. They appreciated the individual freedom they had enjoyed in their youth, especially in contrast to the constraints that had governed their mothers' lives. But, as one woman put it, "a woman, once she grows up, has a family, it's too much to expect her to do everything all by herself." A woman who farmed with her husband in the neighborhood where both had spent their childhood declared, "My generation was the last where we really shared our burdens. Sisters, mothers and in-laws, friends and neighbors—we all knew what was needed and helped each other out without being asked. Today people get together only to socialize, and most of the time everyone's too busy. We were much busier when we were farming, but we did more together—all the time, not just on special occasions."

Farm women agreed that meshing childcare with employment was more difficult than it had been to integrate children into their own work. They had felt under pressure when they had to tend the baby and help get in the hay simultaneously, but they improvised ways to combine their routine responsibilities, taking infants with them to the barn and the fields, having older siblings watch toddlers, and involving school-age children in the work, or at least letting them play at adult chores. When their children's needs were more than they could handle, especially in case of sickness, they could rely both on female relatives and on the men in the household. They concurred that day-care centers were an inadequate substitute—even if childcare did not cost almost as much as many women earned at their jobs. "Women can't earn enough to pay other people to do the things that we used to do for ourselves and each other," one woman stated succinctly.

As these older women reflected on the problems their adult daughters and daughters-in-law had in juggling their family and work responsibilities, they said that the absence of men from the household was as serious as the distance from female friends and relatives. Fathers with city jobs no longer took care of children in the ways they remembered their farmer-fathers doing. "Of course, you *worked*," one emphasized; "you didn't just play. But your father was *there* all the time. And whatever you really needed, you could count on him to do." Today women have to rely mostly on their husbands, "but the man's away at work all day. And when the woman also has an outside job, then it's even worse." "Of course, they have to make ends meet," the older women acknowledged; "today that takes two incomes, just as it used to take both husband and wife working on the farm." But families did not seem to cooperate as much as before.

Women who had grown up after World War I thought that the new emphasis placed by popular culture on heterosexual intimacy had helped bring husbands and wives together. Although their parents had worked side by side like horses in harness, they had not always communicated very well. In the 1920s and 1930s, the idea that spouses should talk over problems helped promote cooperation. But today, in their view, the lack of shared daily work routines undermines mutual understanding. Most women saw rising divorce rates not as evidence that married couples have more problems than they had in the past but rather as a sign of what happens when they are isolated from the support and guidance of their kin and close friends. "There was many a couple separated even then," one woman acknowledged. "But that was only in really serious cases, like infidelity or abuse. Or the man deserted, just up and left his wife and children and disappeared. But the ordinary troubles that beset husbands and wives from time to time—they didn't get so serious." Another explained, "Maybe it was because, on a farm, you couldn't just get a divorce; that would be the end of the farm as well as the family. But maybe it was because, in a community like this, you had help in times of trouble. You weren't alone; your grievances didn't seem so serious. Not that relatives and friends butted in; usually they didn't. But they were always there to help ease your burdens." With more support from others, a third woman explained, "husbands and wives didn't turn on each other so much."

Women who had spent most of their lives in the Nanticoke Valley concurred that since the 1970s women were speaking up and acting for themselves, as they had done during their own youth. Feminism or "women's lib" (as it was then called in the media) seemed to them a new way of saying what had always been obvious—that women deserved to be treated with respect and to be listened to because they did at least half of the world's work. Some said that farm women "had always been liberated." Other Americans had held strange ideas about women's place, but by the early 1980s they caught on to what rural people had always understood. Yet the autonomy that contemporary women enjoyed seemed to them to have come at a high price. Women faced more difficult circumstances once the household and workplace were separated and households were more dependent on the purchase of consumer goods than on shared labor and mutual aid. In rural women's way of thinking, mutuality between husbands and wives and reciprocity between relatives and neighbors were interconnected. The postwar demise of diversified farming, with its gender flexibility, work sharing, and customary cooperation with neighbors, undermined the conditions that these rural women felt had sustained them.

As housework became more distinct from income-producing labor, it was socially devalued. The pride in subsistence production that farm women and men had shared vanished, and rural women were left with routine household chores in better-equipped but increasingly unproductive households. Installing running water, inside bathrooms, furnaces, hot water heaters, and electric or gas appliances, as most rural residents did, was poor compensation for the cold cellars filled with rows of canned goods that women had grown and their pleasure in putting food on the table that they had produced as well as prepared. An older woman said to me in an ironic tone, "What's the use of having all these things to 'lighten my labor' when I no longer have any real work to do in the kitchen?"

Families who had relied on both farming and wage earning now sent all adults into town, except mothers with small children—although some of them did domestic work by the day as well. Two incomes were necessary because most jobs that required no postsecondary education or training paid low wages and offered little security. Experience was important to seniority, but people with interrupted work histories—especially women and farmers— had fewer opportunities for upward mobility in the workplace. Women were employed almost entirely in sex-segregated blue-collar jobs, as stitchers in the shoe or overall factories and assemblers in electronics plants, or in pink-collar jobs, as cashiers, secretaries, teachers, and nurses, where their work was devalued because of gender. Employment in a segregated labor market with sex-linked wage differentials was profoundly different from contributing to the production of milk or eggs, where the value of the commodity rather than the sex of the worker was what counted. As one woman who had worked both on and off the farm pointed out to me while gesturing at a battered milk can in her back kitchen, "nobody knew who milked the cows or hauled the cans to the creamery. They couldn't pay less because it was a woman's work."

Notes

INTRODUCTION

1. For alerting me to the importance of understanding the local histories that residents tell, I am indebted to T. H. Breen, *Imagining the Past: East Hampton Histories* (Reading, Mass.: Addison-Wesley, 1989).

2. For the earlier history of the Nanticoke Valley, see my previous book, Nancy Grey Osterud, *Bonds of Community: The Lives of Farm Women in Nineteenth-Century New York* (Ithaca: Cornell University Press, 1991).

3. Mary Neth, *Preserving the Family Farm: Women, Community, and the Foundations of Agribusiness in the Midwest, 1900–1940* (Baltimore: Johns Hopkins University Press, 1995), 1–2.

4. Hall S. Barron, *Mixed Harvest: The Second Great Transformation in the Rural North, 1870–1930* (Chapel Hill: University of North Carolina Press, 1997).

5. Michael Frisch had the most significant influence on how I approached these oral histories; he probes these issues in *A Shared Authority: Essays on the Craft and Meaning of Oral and Public History* (Albany: SUNY Press, 1990).

6. For more detailed discussion, see Grey Osterud, "Listening for the Contradictions in Rural Women's Life-History Narratives," *Women's History Magazine* (UK) 58 (spring–summer 2008): 4–11. Feminist approaches to the problems inherent in oral history and life-history narratives are presented in three classic collections: Shari Benstock, ed., *The Private Self: Theory and Practice of Women's Autobiographical Writings* (Chapel Hill: University of North Carolina Press, 1988); Personal Narratives Group, ed., *Interpreting Women's Lives: Feminist Theory and Personal Narratives* (Bloomington: Indiana University Press, 1989); Sherna Berger Gluck and Daphne Patai, eds., *Women's Words: The Feminist Practice of Oral History* (New York: Routledge, 1991). Exemplary recent collections include Susan H. Armitage, with Patricia Hart and Karen Weathermon, eds., *Women's Oral History: The Frontiers Reader* (Lincoln: University of Nebraska Press, 2002); Mary Jo Maynes, Jennifer L. Pierce, and Barbara Laslett, *Telling Stories: The Use of Personal Narratives in the Social Sciences and History* (Ithaca: Cornell University Press, 2008). Equally influential in shaping my thinking were British and European analyses: Carolyn Steedman, *Landscape for a Good Woman: A Story of Two Lives* (London: Virago Press, 1986); Carolyn Steedman, *Past Tenses: Essays on Writing, Autobiography, and History* (London: Rivers Oram Press, 1992); Liz Stanley, *The Auto/Biographical I* (Manchester, UK: Manchester University Press, 1992); Luisa Passerini, *Fascism in Popular Memory: The*

Cultural Experience of the Turin Working Class, trans. Robert Lumley and Jude Bloomfield (Cambridge, UK: Cambridge University Press, 1987); Allesandro Portelli, *The Death of Luigi Trastulli, and Other Stories: Form and Meaning in Oral History* (Albany: SUNY Press, 1991); Allesandro Portelli, *The Battle of Villa Giulia: Oral History and the Art of Dialogue* (Madison: University of Wisconsin Press, 1997); Tess Cosslett, Celia Lury, and Penny Summerfield, eds., *Feminism and Autobiography: Texts, Theories, Methods* (London: Routledge, 2000); Selma Leydesdorff, Luisa Passerini, and Paul Thompson, eds., *Gender and Memory,* 2nd ed. (New Brunswick, N.J.: Transaction Press, 2005); Lynn Abrams, *Oral History Theory* (London: Routledge, 2010).

7. Urban dwellers often portrayed rural residents as consumed by petty status rivalries and family grudges. These characteristics were prominent in the most notorious sociological study of a small town in New York State, Arthur J. Vidich and Joseph Bensman, *Small Town in Mass Society: Class, Power, and Religion in a Rural Community* (1958; rev. ed., Princeton: Princeton University Press, 1968). Residents easily identified the place as Owego, located in the next valley west of the Nanticoke Creek. Faculty members at Cornell University, which had sponsored the study, took vehement exception to the book's violation of subjects' privacy, while residents objected angrily to the negative and condescending portrayal of individuals and social relations.

8. Manuscript schedules, Population, United States Census, Towns of Maine and Nanticoke, 1880, 1900, and 1920, in the New York State Library, Albany.

9. Ross McGuire, "Lumbermen, Farmers, and Artisans," in *Working Lives: Broome County, New York, 1800–1930,* ed. Ross McGuire and Nancy Grey Osterud, 1–39 (Binghamton, N.Y.: Roberson Center for the Arts and Sciences, 1982).

10. The largest employer was the Endicott Johnson Shoe Company. See Gerald Zahavi, "Negotiated Loyalty: Welfare Capitalism and the Shoeworkers of Endicott-Johnson, 1920–1940," *Journal of American History* 70 (December 1983): 602–20; Gerald Zahavi, *Workers, Managers, and Welfare Capitalism: The Shoemakers and Tanners of Endicott Johnson, 1890–1950* (Urbana: University of Illinois Press, 1988).

11. See Hal S. Barron, *Those Who Stayed Behind: Rural Society in Nineteenth-Century New England* (New York: Cambridge University Press, 1984).

12. Liberty Hyde Bailey, *The Country Life Movement in the United States* (New York: Macmillan, 1911); William L. Bowers, *The Country Life Movement in America, 1900–1920* (Port Washington, N.Y.: Kennikat Press, 1974); Dennis Roth, "The Country Life Movement," in *Federal Rural Development Policy in the Twentieth Century,* ed. Dennis Roth, Anne B. W. Effland, and Douglas E. Bowers (Washington, D.C.: Economic Research Service, U.S. Department of Agriculture, 2002); David B. Danbom, *The Resisted Revolution: Urban America and the Industrialization of Agriculture, 1900–1930* (Ames: Iowa State University Press, 1979). Paula Baker, *The Moral Frameworks of Public Life: Gender, Politics, and the State in Rural New York, 1870–1930* (New York: Oxford University Press, 1991), places the Country Life Movement in a longer-term political and cultural context and considers the gendered character of rural political activism.

13. Neth, *Preserving the Family Farm,* offers an exemplary social-historical account of the differences between the advocates of capitalist agriculture and organized farmers' movements.

14. This summary is based on E. R. Minns, "Broome County: An Account of Its Agriculture and of Its Farm Bureau," Circular no. 2, Farm Bureau of New York State, 1914; Orville Merton Kile, *The Farm Bureau Movement* (New York: Macmillan, 1921), 92–97; Jasper F. Eastman, "Agriculture," in *Binghamton and Broome County, New York: A History,* ed. William Foote Seward, 134–49 (New York and Chicago: Lewis Historical Publishing Co., 1924); New York State Farm Bureau Federation, "The Farm Bureau Carries On: The Growth of an Idea," Ithaca, ca. 1945; Orville Merton Kile, *The Farm Bureau through Three Decades* (Baltimore: Waverly Press, 1948), 27–46.

15. For a vivid account of the conflict in a nearby community written by a former Dairymen's League organizer, see Edward R. Eastman, *The Trouble Maker* (New York: Macmillan, 1925).

16. *Broome County Farm and Home Bureau News,* April 1923, 2, in Broome County Cooperative Extension Archives, Binghamton, N.Y.

17. Individuals' places of birth were recorded by the U.S. and New York State censuses, but the categories used range from broad political units, such as "Austria-Hungary," to specific regions, such as "Galicia" and "Bohemia." In 1925, according to the manuscript schedules of the New York State Census, the birthplaces of the largest numbers of foreign-born individuals living in the Nanticoke Valley were Poland and Galatia [sic] (70); Austria and Austria-Hungary (65); and Czechoslovakia and Bohemia (24). Smaller but still significant numbers came from Germany, Russia and the Ukraine, and Armenia.

18. Manuscript schedules, Population, U.S. Census, 1920, Towns of Maine and Nanticoke; Manuscript schedules, Population, New York State Census, 1925, Towns of Maine and Nanticoke; both in the New York State Library, Albany.

19. *Sixteenth Census of the United States, 1940, Agriculture,* Vol. 1, Pt. 1 (Washington, D.C.: U.S. Government Printing Office, 1942), 264 (table 9).

20. Ralph Young, diaries, 1913–1945, held by the family.

21. The allusion is to Benedict Anderson, *The Imagined Community: Reflections on the Origin and Spread of Nationalism* (1983; rev. ed. London: Verso, 1991). This view offers a corrective to the assumptions about "the community" that shape everyday thinking and have been embedded in social theories that contrast a putatively homogeneous, organic, and united rural community with a diverse, impersonal, and conflict-ridden urban society.

22. Dwight Sanderson, "Community Halls," *Broome County Farm and Home Bureau News,* September 1921, 1.

23. James H. Madison, "Reformers and the Rural Church, 1900–1950," *Journal of American History* 73 (December 1986): 645–68.

24. David B. Danbom, "Rural Educational Reform and the Country Life Movement, 1900–1920," *Agricultural History* 53 (spring 1979): 462–74.

25. Throughout this book, the people I interviewed are identified by pseudonyms that reflect their ethnocultural heritage. Only those who appear in the documentary records I cite or quote, such as members of the Young and McGregor families, are identified by their real names. Transcripts and summaries of the interviews are in the archives of the Nanticoke Valley Historical Society in Maine, New York. Although all the individuals I interviewed have died, these materials remain closed to researchers until everyone mentioned in them is deceased.

CHAPTER 1

1. Mary E. Wilkins [Freeman], "The Revolt of 'Mother,'" *Harpers' Monthly,* September 1890, reprinted in *A New England Nun and Other Stories* (New York: Harper & Brothers, 1891). See Leah Blatt Glasser, *In a Closet Hidden: The Life and Work of Mary E. Wilkins Freeman* (Amherst: University of Massachusetts Press, 1996).

2. For evidence of the story's wide circulation and its consistency with other turn-of-the-century commentaries on rural women's situation, see Ellen Gruber Garvey, "Less Work for 'Mother': Rural Readers, Farm Papers, and the Makeover of 'The Revolt of "Mother,"'" *Legacy* 26, no. 1 (November 2009): 119–49, especially n. 8.

3. Mary E. Wilkins Freeman, "An Autobiography," *Saturday Evening Post,* December 8, 1917.

4. For a strong statement of this viewpoint, see Michael Grimwood, "Architecture and Autobiography in 'The Revolt of "Mother,"'" *American Literary Realism* 40, no. 1 (2008): 66–82.

5. Elizabeth Hampsten considered this striking quality of women's recollections of their childhood in *Read This Only to Yourself: The Private Writings of Midwestern Women, 1880–1910* (Bloomington: Indiana University Press, 1982), *Settlers' Children: Growing Up on the Great Plains* (Norman: University of Oklahoma Press, 1991), and especially her edition of Pauline Neher Diede, *Homesteading on the Knife River Prairies* (Bismarck, N.D.: Germans from Russia Heritage Society, 1983).

CHAPTER 2

1. On the significance of the formal structure of women's life-history narratives, see Lynn Abrams, *Oral History Theory* (London: Routledge, 2010).

2. For more in-depth discussion of my approach to interpreting these problematic aspects of oral histories, see Grey Osterud, "Listening for the Contradictions in Rural Women's Life-History Narratives," *Women's History Magazine* (UK) 58 (spring–summer 2008): 4–11.

3. Carl L. Becker, "Everyman His Own Historian," 1931 Presidential Address to the American Historical Association, *American Historical Review* 37, no. 2 (1932): 221–36, www. historians.org/info/AHA_history/clbecker.htm (accessed May 28, 2009). Becker's dictum has been used more often to critique historical knowledge as shaped by historians' present preoccupations than to point us toward the interpretive work that ordinary people do. His democratic "Mr. Everyman" is aware of the past and uses it to lend meaning to his actions in the present, although the usable past is memory rather than reconstructed history.

4. In thinking through these questions, I have been enlightened by recent works on women's history outside the United States, especially Lynn Abrams, *Myth and Materiality in a Woman's World: Shetland, 1800–2000* (Manchester, UK: Manchester University Press, 2005); Katherine Holden, *The Shadow of Marriage: Singleness in England, 1914–60* (Manchester, UK: Manchester University Press, 2007); Mary Jo Maynes, Jennifer L. Pierce, and Barbara Laslett, *Telling Stories: The Use of Personal Narratives in the Social Sciences and History* (Ithaca: Cornell University Press, 2008).

5. For an especially sensitive and comprehensive discussion of these issues in interpreting oral histories, see Abrams, *Oral History Theory*.

6. The major study of women's landholding is Anne B. Effland, Denise M. Rodgers, and Valerie Grim, "Women as Agricultural Landowners: What Do We Know about Them?" *Agricultural History* 67 (spring 1993): 235–61. After a careful examination of federal government surveys of farm owners and numerous regional studies, they concluded that these statistical analyses reveal little about women's landownership because they lump together farms that were jointly owned by husband and wife with farms that were held solely by the husband. The presumption that adult men exercised authority regardless of joint or individual ownership blurs the actual legal situation. Still, Effland, Rodgers, and Grim offered some reliable data showing that a substantial proportion of farm owners have been women, both jointly and in their own names.

7. A popular Midwestern periodical for rural women was called *The Farmer's Wife*. Proponents of agribusiness, home economics, and Progressive reform defined farm women ideologically as reproducers rather than producers and pushed them to conform to urban middle-class norms of domesticity. Many studies of rural women's participation in farm organizations have shown that women resisted these efforts both individually and collectively and continued to act and think of themselves as producers. See Kathleen R. Babbitt, "The Productive Farm Woman and the Extension Home Economist in New York State, 1920–1940," *Agricultural History* 67 (spring 1993): 83–101; Paula Baker, *The Moral Frameworks of Public Life: Gender, Politics, and the State in Rural New York, 1870–1930* (New York: Oxford University Press, 1991); Lu Ann Jones, *Mama Learned Us to Work: Farm Women in the New South* (Chapel Hill: University of North Carolina Press, 2002); Virginia E. McCormick, ed., *Farm Wife: A Self-Portrait* (Ames: Iowa State University Press, 1990); Sally McMurry, *Families and Farmhouses in Nineteenth-Century America: Vernacular Design and Social Change*, rev. ed. (Knoxville: University of Tennessee Press, 1997); Debra Ann Reid, *Reaping a Greater Harvest: African Americans, the Extension Service, and Rural Reform in Jim Crow Texas* (College Station: Texas A & M University Press, 2007).

8. When I conducted most of these interviews, the shift to full-time employment for married women, even those with young children and elderly parents to care for, was very much on these older women's minds. Their adult daughters and daughters-in-law, who held paid jobs outside the household, were juggling competing responsibilities they felt they had been able to combine.

9. For a comparative study of inheritance patterns, see Sonya Salamon, *Prairie Patrimony: Family, Farming, and Community in the Midwest* (Chapel Hill: University of North Carolina Press, 1992); Sonya Salamon and Karen Davis-Brown, "Farm Continuity and Female Land Inheritance: A Family Dilemma," in *Women and Farming: Changing Roles, Changing Structures*, ed. Wava Haney and Jane B. Knowles, 195–210 (Boulder, Colo.: Westview Press, 1979).

10. Hilda had died at the age of seventy-nine, ten years before I interviewed Carrie.

11. Joint property ownership prevented a husband from selling or mortgaging land without a woman's consent. In New York State, laws protecting the property rights of married women attempted to ensure that women could not be coerced into agreeing to the sale of the family farm, but they were difficult to enforce in the absence of joint title. In the event of divorce, a joint owner would be entitled to a property settlement as long as she was not at fault. Most women who inherited a farm when their husband died could make a will designating their own heirs, although some couples made arrangements that obligated an adult child to support them in old age in exchange for inheriting the land after their decease. A widow who was the joint owner of the family farm was not necessarily able to take over the enterprise without the estate going through probate, however, which created serious practical difficulties until legal reforms in the 1980s.

12. When I interviewed Amanda Thompson, she had lived out her adulthood as a single woman. She surprised herself as well as others when, at the age of eighty-three, she married a widower she had known for years.

CHAPTER 3

1. Hugh L. Cosline, "Buying a Farm on a Small Capital: How a Broome County Farmer Paid for His Farm under Difficulties," *American Agriculturist*, August 14, 1926, 3, 13.

2. Ralph Young, unpublished memoir. Ralph Young, who was eighty when I interviewed him and his daughter, Margaret Young Coles, allowed me to study the diaries he had kept from the age of eighteen. The diaries and memoir belong to the family and, like the interviews, are quoted with the permission of Ralph Young and Margaret Young Coles.

3. This phrase was used during the late 1910s and 1920s by agricultural experts and rural sociologists who sought to counteract the acknowledged tendency of adult men in farm families to disregard the opinions of their wives and children. Both agricultural extension programs and the Farm and Home Bureau advocated holding family meetings in which all family members discussed important issues involving the farm. I heard this phrase in interviews with members of several Nanticoke Valley families who thought of themselves as progressive farmers, especially those who operated in partnerships between parents and adult children.

4. Cosline, "Buying a Farm on a Small Capital," 3.

5. Ibid., 13.

6. Ibid.

7. Ibid.

8. See Nancy Grey Osterud, *Bonds of Community: The Lives of Farm Women in Nineteenth-Century New York* (Ithaca: Cornell University Press, 1992), 53–56, 64–65, 80–81, 92, 94, 104, 114–18, 120–21, 131, 139–40, 160–68, 218–20, 231–32, 237–40, 246, 253–54, 257, 267–68, photographs on 44–47. The close resemblance between the Riley and the Young families is striking. Although George Riley and Ralph Young were more than a generation apart and were not even distantly related, they exhibited similar patterns of gender integration in work and sociability and deep commitments to community cooperation.

9. These patterns were very similar to those found in the late nineteenth century; see Osterud, *Bonds of Community,* 231–54, 262–74.

CHAPTER 4

1. Hugh L. Cosline, "Buying a Farm on a Small Capital: How a Broome County Farmer Paid for His Farm under Difficulties," *American Agriculturist*, August 14, 1926, 3.

2. Histories of American agriculture in the twentieth century include Jane H. Adams, *The Transformation of Rural Life: Southern Illinois, 1890–1990* (Chapel Hill: University of North Carolina Press, 1994); Willard W. Cochrane, *The Development of American Agriculture: A Historical Analysis* (Minneapolis: University of Minnesota Press, 1993); Paul K. Conkin, *A Revolution Down on the Farm: The Transformation of American Agriculture since 1929* (Lexington: University Press of Kentucky, 2009); David B. Danbom, *Born in the Country: A History of Rural America* (Baltimore: Johns Hopkins University Press, 2006); Gilbert C. Fite, *American Farmers: The New Minority* (Bloomington: Indiana University Press, 1981); Deborah Fitzgerald, *Every Farm a Factory: The Industrial Ideal in American Agriculture* (New Haven, Conn.: Yale University Press, 2003); Bruce L. Gardner, *American Agriculture in the Twentieth Century: How It Flourished and What It Cost* (Cambridge, Mass.: Harvard University Press, 2006); R. Douglas Hurt, *Problems of Plenty: The American Farmer in the Twentieth Century* (Chicago: Ivan Dee, 2002); Katherine Jellison, *Entitled to Power: Farm Women and Technology, 1913–1963* (Chapel Hill: University of North Carolina Press, 1993); Mary Neth, *Preserving the Family Farm: Women, Community, and the Foundations of Agribusiness in the Midwest, 1900–1940* (Baltimore: Johns Hopkins University Press, 1995). For a summary of recent trends in off-farm wage-earning, see Rachel Ann Rosenfeld, *Farm Women: Work, Farm, and Family in the United States* (Chapel Hill: University of North Carolina Press, 1985), 141–85. Most of the scholarship on "urban fringe farming" focuses on the loss of farmland to housing and other forms of development and on conflicts over taxation, land use, and environmental issues between farmers and suburbanites rather than on rural residents' combination of farming with wage-earning.

3. Changes in the census's definition of a farm between 1900 and 1940 involved only the very smallest farms and do not compromise the comparability of the data cited here. The census's bias toward regarding married women as engaged in housekeeping rather than farming, however, meant that the small-scale, subsistence-oriented operations carried on by women in rural and urban communities alike—cultivating large vegetable gardens, keeping small poultry flocks producing chickens and eggs, and even pasturing a milk cow—were deliberately excluded.

4. See Nancy Grey Osterud, *Bonds of Community: The Lives of Farm Women in Nineteenth-Century New York* (Ithaca: Cornell University Press, 1991). See also Grey Osterud, "Farm Crisis and Rural Revitalization in South-Central New York during the Early Twentieth Century," *Agricultural History* 84 (2010): 141–65.

5. See Lawrence Bothwell, *Broome County Heritage* (Woodland Hills, Calif.: Windsor Publications, 1983); Ross McGuire, "Lumbermen, Farmers, and Artisans: The Rural Economy of the Nineteenth Century," in *Working Lives: Broome County, New York, 1900–1930,* ed. Ross McGuire and Nancy Grey Osterud, 1–39 (Binghamton, N.Y.: Roberson Center for the Arts and Sciences, 1982).

6. See Gerald Zahavi, "Negotiated Loyalty: Welfare Capitalism and the Shoeworkers of Endicott-Johnson, 1920–1940," *Journal of American History* 70 (December 1983): 602–20; Gerald Zahavi, *Workers, Managers, and Welfare Capitalism: The Shoemakers and Tanners of Endicott Johnson, 1890–1950* (Urbana: University of Illinois Press, 1988).

7. Theodore Roosevelt, "The Abandoned Farm," *Century* 97 (April 1911): 938–41.

8. Liberty Hyde Bailey, *The Country Life Movement in the United States* (New York: Macmillan, 1911). See also William L. Bowers, *The Country Life Movement in America, 1900–1920* (Port Washington, N.Y.: Kennikat Press, 1974); Dennis Roth, "The Country Life Movement," in *Federal Rural Development Policy in the Twentieth Century,* ed. Dennis Roth, Anne B. W. Effland, and Douglas E. Bowers (Washington, D.C.: Economic Research Service, U.S. Department of Agriculture, 2002); Roy V. Scott, *The Reluctant Farmer: The Rise of Agricultural Extension to 1914* (Urbana: University of Illinois Press, 1978); David B. Danbom, *The Resisted Revolution: Urban America and the Industrialization of Agriculture, 1900–1930* (Ames: Iowa State University Press, 1979).

9. Edward G. Misner, "Milk Production Costs," *Journal of Farm Economics* 1, no. 3 (October 1910): 97–101, quotations on 97, 101. See also Lawrence M. Vaughan, "Abandoned Farm Areas in New York," Cornell University Agricultural Experiment Station Bulletin no. 490, Cornell University, July 1929, 1–285, esp. 250–51.

10. *Fourteenth Census of the United States, 1920, Agriculture,* Vol. 5 (Washington, D.C.: U.S. Government Printing Office, 1921), Appendix B, 919.

11. Vaughan, "Abandoned Farm Areas in New York," quotations on 9–10, statistics on 250.

12. Improved land included "all land regularly tilled or mowed, land pastured and cropped in rotation, land in gardens, orchards, vineyards and nurseries, and land occupied by farm buildings." *Thirteenth Census of the United States, 1910, Agriculture,* Vol. 5 (Washington, D.C.: U.S. Government Printing Office, 1911), 25. The problem with this definition is that it did not specify how permanent pasture was to be classified. In 1920, the instructions to enumerators specified that permanent pasture should be included in improved land. *Fourteenth Census of the United States, 1920, Agriculture,* Vol. 5, 17. Old fields that had been left fallow, however, were probably included in unimproved land.

13. "The Agricultural Progress of Fifty Years," *Twelfth Census of the United States, 1900, Agriculture,* Vol. 5 (Washington, D.C.: U.S. Government Printing Office, 1911), xxii.

14. For a biography that places Warren at the forefront of the development of agricultural economics, see Bernard F. ("Bud") Stanton, *George F. Warren: Farm Economist* (Ithaca: Department of Applied Economics and Management, Cornell University, 2008).

15. These terms were defined very carefully to be parallel to those used in other businesses but take into account the distinctive position of farmers. *Expenses* included the value of the labor of unpaid family workers, but not household and personal expenditures. *Net income* represented the difference between receipts and expenses. *Labor income* was computed by subtracting a 5 percent return on investment from net income; *return on investment,* conversely, was computed by subtracting the value of the farmer's labor from net income.

16. *Specialized farms* were defined as those that derived 40 percent of their income from a single operation.

17. Maurice C. Burritt, "The Incomes of 178 New York Farms," Cornell University Agricultural Experiment Station Bulletin no. 271, Cornell University, December 1909, 15–27.

18. George F. Warren and Kenneth C. Livermore, "An Agricultural Survey: Townships of Ithaca, Dryden, Danby, and Lansing, Tompkins Country, New York," Cornell University Agricultural Experiment Station Bulletin no. 295, Cornell University, March 1911, 377–569, quotations on 431–34, 395–97, 389, 441, 444.

19. Ibid., 400, 556. Because women were seldom hired to work on farms, male farmers had no wage scale on which to base these estimates. The estimate that family members did an average of 70 percent of the labor is more accurate, but may well underestimate the amount of time women spent in the barn and the fields.

20. Warren and Livermore, "Agricultural Survey," 508–9, 544–48, 562. Remarkably, the survey paid special attention to women farmers. Most female landowners were wives and daughters who had inherited land rather than embarking on farming by themselves. Half rented out their farms, but half lived on them, sometimes with tenants who operated the farm on shares. The 15 percent of female landholders who actively managed their own farms made more money than the rest. Detailed records demonstrated that women who chose to farm might do so profitably. Finally, the authors recognized that some farmers held off-farm jobs. These families produced their own living and some cash income from sales. Indeed, their farm labor incomes compared favorably with the poorest third of full-time farmers.

21. George F. Warren, "Some Important Factors for Success in General Farming and in Dairy Farming," Cornell University Agricultural Experiment Station Bulletin no. 248, Cornell University, July 1914, 677.

22. Arthur L. Thompson, "The Cost of Producing Milk on 174 Farms in Delaware County, New York," Cornell University Agricultural Experiment Station Bulletin no. 364, Cornell University, October 1915, 128, 142, 151, 158.

23. Edward G. Misner, "An Economic Study of Dairying on 149 Farms in Broome County, New York," Cornell University Agricultural Experiment Station Bulletin no. 409, Cornell University, April 1922, 273–443, quotations on 277. Of the farms studied, 50 were in the river valleys and 99 in the uplands. A close look at Misner's map (on 278) suggests that between 12 and 22 of these upland farms were in the Nanticoke Valley—which, in Misner's terms, is almost entirely an upland area.

24. Ibid., 287.

25. Ibid., 321, 326, 338, 350.

26. Ibid., 346–47, 358–59.

27. Edward G. Misner, "An Economic Study of Dairying on 163 Farms in Herkimer County, New York," Cornell University Agricultural Experiment Station Bulletin no. 432, Cornell University, September 1924, 1–59.

28. Ibid., data on 15, 16, conclusions on 38–39, 41, 49.

29. P. H. Stephens, "Economic Studies of Dairy Farming in New York, XI. Success in Management of Dairy Farms as Affected by the Proportion of the Factors of Production," Cornell University Agricultural Experiment Station Bulletin no. 562, Cornell University, June 1933, 1–45, quotation on 44.

30. *Twelfth Census of the United States, 1900, Agriculture*, Vol. 5, ccxxiv.

31. Johann C. Neethling, "Economic Studies of Dairy Farming in New York State, IX. Grade B Milk with Cash Crops and Mixed Hay Roughage," Cornell University Agricultural Experiment Station Bulletin no. 483, Cornell University, June 1929, 1–93, quotations on 92–93. See also Edward G. Misner and A. T. M. Lee, "Economic Studies of Poultry Farming in New York, I. Commercial Poultry Farms, 1926, 1929, 1930, 1931, 1932, 1933," Cornell University Agricultural Experiment Station Bulletin no. 684, Cornell University, December 1937, 1–122.

32. *Fifteenth Census of the United States, 1930, Agriculture*, Vol. 4 (Washington, D.C.: U.S. Government Printing Office, 1931), Appendix, 1–7. To be defined as "part-time," farms also could not yield more than $750 worth of produce per year.

33. Changes in the classification scheme adopted in 1940, particularly the elimination of "general" farms and the identification of all farms on the basis of their largest product, meant that smaller-scale dairy operations were now included, which we would expect to lead to an increase in their number. That the opposite trend occurred testifies to the strength of the actual shift.

34. Gould Coleman and Sarah Elbert, "Farming Families: The Farm Needs Everyone," *Research in Rural Sociology and Development* 1 (1984): 61–78.

35. In the terms of capitalist agriculture as described by a group of Canadian economists, they were propertied laborers who engaged in self-exploitation; see John Emmeus Davis, "Capitalist Development and the Exploitation of the Propertied Labourer in the Era of Wage Labour," *Comparative Studies in Society and History* 20 (1980): 545–86.

36. For an astute analysis of the extent to which farmers' cooperatives came to resemble agribusiness corporations, see Hal S. Barron, "Bringing Forth Strife: The Ironies of Dairy Organization in the New York Milkshed," in *Mixed Harvest: The Second Great Transformation in the Rural North, 1870–1930* (Chapel Hill: University of North Carolina Press, 1997), 81–105.

CHAPTER 5

1. Nancy Grey Osterud, *Bonds of Community: The Lives of Farm Women in Nineteenth-Century New York* (Ithaca: Cornell University Press, 1991).

2. The high degree of cooperation and flexibility that characterized gender relations in dairy farming families was first remarked in Arthur F. Raper, "The Dairy Areas," in *Rural Life in the United States*, ed. Carl C. Taylor, Arthur F. Raper, Douglas Ensminger, et al., 414–36 (New York: Knopf, 1949), esp. 418.

3. For an overview, see Nancy Grey Osterud, "Gender and the Transition to Capitalism in Rural America," in *American Rural and Farm Women in Historical Perspective*, ed. Joan M. Jensen and Nancy Grey Osterud, 14–29 (Berkeley: University of California Press, 1994).

4. Foundational works on rural women's history in North America include Wava G. Haney and Jane B. Knowles, eds., *Women and Farming: Changing Roles, Changing Structures* (Boulder, Colo.: Westview Press, 1988); Joan M. Jensen, ed., *With These Hands: Women Working on the Land* (Old Westbury, N.Y.: Feminist Press and McGraw-Hill Book Co., 1981); Joan M. Jensen, *Loosening the Bonds: Mid-Atlantic Farm Women, 1750–1850* (New Haven, Conn.: Yale University Press, 1986); Joan M. Jensen, *Promise to the Land: Essays on Rural Women* (Albuquerque: University of New Mexico Press, 1991); Joan M. Jensen, *Calling This Place Home: Women on the Wisconsin Frontier, 1850–1925* (Minneapolis: Minnesota Historical Society Press, 2006); Seena B. Kohl, *Working Together: Women and Family in Southwestern Saskatchewan* (Toronto: Holt, Rinehart and Winston of Canada, 1976).

5. Mary Neth, *Preserving the Family Farm: Women, Community, and the Foundations of Agribusiness in the Midwest, 1900–1940* (Baltimore: Johns Hopkins University Press, 1995), 1–2.

6. The contrast is especially clear in two ethnographic studies by Deborah Fink, an anthropologist. *Open Country, Iowa: Rural Women, Tradition, and Change* (Albany: SUNY Press, 1986) portrays a densely settled, fertile rural landscape in which women had access to significant material and social resources. In contrast, *Agrarian Women: Wives and Mothers in Rural Nebraska* (Chapel Hill: University of North Carolina Press, 1992) depicts a sparsely settled, resource-poor region in which women were isolated and subordinated to men. Comparing gender relations in these two rural societies suggests that women's power depended not only on their participation in productive labor but also on the strength of their kin connections and the long-term stability of the rural population, which nurtured social ties.

7. Neth, *Preserving the Family Farm*, 32–33. The best explanation of what social scientists call a bargaining approach to the analysis of unequal power relations in farming families comes from an Indian economist; see Bina Agarwal, *A Field of One's Own: Gender and Land Rights in South Asia* (Cambridge, UK: Cambridge University Press, 1994), esp. 51–81.

8. Notable historical and sociological works that analyze the division of labor relative to the scale and degree of specialization of the farm and the sex and age composition of the household include Jane H. Adams, *The Transformation of Rural Life: Southern Illinois, 1890–1990* (Chapel Hill: University of North Carolina Press, 1994); Peggy F. Barlett, *American Dreams, Rural Realities: Family Farms in Crisis* (Chapel Hill: University of North Carolina Press, 1993); Louise I. Carbert, *Agrarian Feminism: The Politics of Ontario Farm Women* (Toronto: University of Toronto Press, 1995); Marjorie Griffin Cohen, *Women's Work, Markets, and Economic Development in Nineteenth-Century Ontario* (Toronto: University of Toronto Press, 1988); Gould Coleman and Sarah Elbert, "Farming Families: The Farm Needs Everyone," *Research in Rural Sociology and Development* 1 (1984): 61–78; Milton Coughenor and Louis E. Swanson, "Work Statuses and Occupations of Men and Women in Families and the Structure of Farms," *Rural Sociology* 48 (spring 1983): 23–43; Sarah Elbert, "The Farmer Takes a Wife: Women in American Farming Families," in *Women, Households, and the Economy*, ed. Lourdes Beneria and Catherine R. Stimpson, 172–97 (New Brunswick, N.J.: Rutgers University Press, 1987); Sarah Elbert, "Women and Farming: Changing Structures, Changing Roles," in Haney and Knowles, *Women and Farming*, 245–64; Lorraine Garkovich and Janet Bokemeier, "Agricultural Mechanization and American Farm Women's Economic Roles," in Haney and Knowles, *Women and Farming*, 211–28; Barbara Handy-Marcello, *Women of the Northern Plains: Gender and Settlement on the Homestead Frontier, 1870–1930* (Minneapolis: Minnesota Historical Society Press, 2005); Katherine Jellison, *Entitled to Power: Farm Women and Technology, 1913–1963* (Chapel Hill: University of North Carolina Press, 1993); Sally McMurry, *Transforming Rural Life: Dairying Families and Agricultural Change, 1820–1885* (Baltimore: Johns Hopkins University Press, 1995); Rachel Ann Rosenfeld, *Farm Women: Work, Farm, and Family in the United States* (Chapel Hill: University of North Carolina Press, 1985); Carolyn Sachs, *The Invisible Farmers: Women in Agricultural Production* (Totowa, N.J.: Rowman & Allanheld, 1983); Sonya Salamon, *Prairie Patrimony: Family, Farming, and Community in the Midwest* (Chapel Hill: University of North Carolina Press, 1992); Rebecca Sharpless, *Fertile Ground, Narrow Choices: Women on*

Texas Cotton Farms, 1900–1940 (Chapel Hill: University of North Carolina Press, 1999); Melissa Walker, *"All We Knew Was to Farm": Rural Women in the Upcountry South, 1919–1941* (Baltimore: Johns Hopkins University Press, 2000).

9. Jeanne Boydston, *Home and Work: Housework, Wages, and the Ideology of Labor in the Early Republic* (New York: Oxford University Press, 1990), locates this shift in the dominant culture during the early nineteenth century, but for many rural Americans it was not apparent until the early twentieth century.

CHAPTER 6

1. Similar patterns have been found by other historians of agriculture who have looked at women's work in other regions as farming was transformed by capitalist agribusiness during the twentieth century. See, for example, Jane H. Adams, *The Transformation of Rural Life: Southern Illinois, 1890–1990* (Chapel Hill: University of North Carolina Press, 1994); Sarah Elbert, "The Farmer Takes a Wife: Women in American Farming Families," in *Women, Households, and the Economy,* ed. Lourdes Beneria and Catherine R. Stimpson, 172–97 (New Brunswick, N.J.: Rutgers University Press, 1987); Deborah Fink, *Open Country, Iowa: Rural Women, Tradition, and Change* (Albany: SUNY Press, 1986); Deborah Fink, *Agrarian Women: Wives and Mothers in Rural Nebraska* (Chapel Hill: University of North Carolina Press, 1992); Katherine Jellison, *Entitled to Power: Farm Women and Technology, 1913–1963* (Chapel Hill: University of North Carolina Press, 1993); Seena B. Kohl, *Working Together: Women and Family in Southwestern Saskatchewan* (Toronto: Holt, Rinehart and Winston of Canada, 1976); Sally McMurry, *Transforming Rural Life: Dairying Families and Agricultural Change, 1820– 1885* (Baltimore: Johns Hopkins University Press, 1995); Mary Neth, *Preserving the Family Farm: Women, Community, and the Foundations of Agribusiness in the Midwest, 1900–1940* (Baltimore: Johns Hopkins University Press, 1995); Sonya Salamon, *Prairie Patrimony: Family, Farming, and Community in the Midwest* (Chapel Hill: University of North Carolina Press, 1992); Rebecca Sharpless, *Fertile Ground, Narrow Choices: Women on Texas Cotton Farms, 1900–1940* (Chapel Hill: University of North Carolina Press, 1999); Melissa Walker, *"All We Knew Was to Farm": Rural Women in the Upcountry South, 1919–1941* (Baltimore: Johns Hopkins University Press, 2000).

2. "V. C. McGregor," *American Agriculturist,* February 2, 1935, 69. See also "McGregor Farms: Successful Family Enterprise," *American Agriculturist,* August 27, 1938, 501.

3. For a perceptive analysis of patterned differences in the ways that women and men perceived and discussed women's farm work, see Shauna Scott, "Drudges, Helpers and Team Players in Oral Historical Accounts of Farm Work in Appalachian Kentucky," *Rural Sociology* 61, no. 2 (1990): 209–26.

4. Ruth Tibbits's mother was Clara's sister. Both her parents were killed when their car was hit by a train around 1910. Ruth's older siblings insisted she keep their family name, so the McGregors did not formally adopt her. She married a young man who worked at the McGregors.

5. I am grateful to Garth, Margaret, Dane, and Ruth McGregor for sharing this diary with me and allowing me to analyze and quote from it here. The diary remains in the family's possession.

6. Men in the Young and Smith families also helped in the house more often than most. In other households, men might carry water on washday but rarely helped make meals or do the dishes.

7. Genealogical research shows that the two families had intermarried in the mid-nineteenth century as well, but by the early twentieth century they were apparently unaware of their distant kinship.

CHAPTER 7

1. Milotice is in the South Moravian region, close to the Slovak border. *Slavish* is an alternate form of *Slavic,* not a synonym for *Slovak.*

2. Interviews with Mrs. Simon Fenson by Nettie Politylo, conducted at her home, 2121 Farm to Market Road, RD #2, in Johnson City, New York, on April 25 and June 20, 1978, for the Broome County Oral History Project, sponsored by Action for Older Persons, Inc., in Binghamton, New York. Tape and unedited transcript in the Broome Country Historical Society Library, Binghamton, New York. Nadya was born in a small town called Wisoko near Brodi, which is now in the western Ukraine; she grew up speaking both Russian and Polish.

CHAPTER 8

1. I conducted initial interviews with a number of women who felt bitter about how their lives had turned out, but in all these cases, the women expressed reluctance to release the tapes and transcripts and I decided not to conduct follow-up interviews. In our off-the-record discussions after I gave them a transcript of our initial conversation, these women expressed concern not about what others might think of them and their families but, rather, about what they thought of themselves and the ways in which they had coped—or had not coped effectively— with the difficult interpersonal situations they had encountered. I did not continue interviewing them because most appeared to be clinically depressed. Although what psychologists call "life review" can be beneficial to some older women with unresolved internal conflicts about the past, in these cases I was more aware of the difference between a historical researcher and a counseling professional.

2. In Broome County, as the agricultural census data reveal, farms that were rented to cash tenants were, on average, larger in scale and more specialized than owner-occupied farms.

3. The most emotionally expressive portions of Clarice's history of the Lanes concern not her own family but that of her father's younger brother, whose parents had left him behind when they went their separate ways when he was very young. He married in 1924, the same year that Sampson Harwood Lane married Maggie Saunders. His wife "couldn't take not knowing where the next meal was coming from, not having wood to keep warm or to cook with, drinking, and a baby each year." Strikingly, Clarice did not blame this woman for leaving her husband and taking only her oldest child with her. After she left, the youngest children were put up for adoption. As an adult, one of those children returned to the Nanticoke Valley and got back in contact with his mother. But meeting his long-lost relatives did not repair the rupture in his life. This sad story emphasizes the lifelong consequences for a child who was left behind by a desperate mother and abandoned by his father and extended family, a fate that Clarice felt fortunate to avoid.

CHAPTER 9

1. For the Nanticoke Valley, see Nancy Grey Osterud, *Bonds of Community: The Lives of Farm Women in Nineteenth-Century New York* (Ithaca: Cornell University Press, 1991), 254–62. For the local, state, and national history of the Grange, see Broome County Granges, Patrons of Husbandry, *Official Directory and History of the Broome County Granges, Patrons of Husbandry* (Binghamton, N.Y.: Patrons of Husbandry, 1941); Leonard Allen, *History of the New York State Grange* (Watertown, N.Y.: Hungerford-Holbrook, 1934); Charles M. Gardner, *The Grange—Friend of the Farmer, 1867–1947* (Washington, D.C.: The National Grange, 1949). D. Sven Nordin, *Rich Harvest: A History of the Grange, 1867–1900* (Jackson: University Press of Mississippi, 1974), recognizes the distinct character of the second phase of the Grange movement, which was centered in the Northeast rather than the Midwest, but overemphasizes the programmatic shift from economic cooperation to education and sociability. Donald B. Marti, *Women of the Grange: Mutuality and Sisterhood in Rural America, 1866–1920* (Westport, Conn.: Greenwood Press, 1991), offers a comprehensive and balanced account of the group's ideas and practices regarding women.

2. On the history of the Dairymen's League, see R. D. Cooper, "Origin and Development of the Dairymen's League," manuscript, ca. 1938, in the Guide to the Dairymen's League Cooperative Association Records, Division of Rare and Manuscript Collections, Cornell University Library. For the organization of the Farm Bureau, the Dairymen's League, and the Grange League Federation in Broome County, see Jasper F. Eastman, "Agriculture," in *Binghamton and Broome County, New York: A History,* ed. William Foote Seward, 134–49 (New York and Chicago: Lewis Historical Publishing, 1924). Eastman's radical views were toned down a bit for this edited volume.

3. Gould P. Coleman, *Education and Agriculture: A History of the New York State College of Agriculture at Cornell University* (Ithaca: Cornell University Press, 1962), although celebrating extension programs' democratization of knowledge, recognizes the capitalist orientation of its advice to farmers.

4. Arthur Raper, a pioneering rural sociologist, noticed dairy farmers' propensity to organize cooperatives; see Arthur F. Raper, "The Dairy Areas," in *Rural Life in the United States,* ed. Carl C. Taylor, Arthur F. Raper, Douglas Ensminger, et al., 414–36 (New York: Knopf, 1949), esp. 424.

5. This line of argument was first developed in Mary Neth, *Preserving the Family Farm: Women, Community, and the Foundations of Agribusiness in the Midwest, 1900–1940* (Baltimore: Johns Hopkins University Press, 1995).

6. The most precise statements of the claim that Broome County farmers founded the first grassroots organization to support a county agent are in Orville Merton Kile's 1921 and 1948 histories of the Farm Bureau. Agricultural experts had offered advice to farmers before 1911; extension agents were employed by state agricultural colleges and experiment stations, and the USDA employed demonstration agents, first in the South under Dr. Seaman A. Knapp and then in Pennsylvania under Dr. William J. Spillman. But "the credit for the first farm bureau belongs to Broome County" because there "the principle of local control and local responsibility was established." Orville Milton Kile, *The Farm Bureau Movement* (New York: Macmillan, 1921), 92–97, quotation on 96. John H. Barron "is considered to be the first 'farm bureau' representative in the U.S." because he was guided by a formal organization of farmers. The "county agent idea" spread rapidly, and other places "followed the pattern of greater local farmer responsibility" initiated by Broome County. Orville Merton Kile, *The Farm Bureau through Three Decades* (Baltimore: Waverly Press, 1948), 27–46, quotations on 32, 41. Nancy K. Berlage, "Organizing the Farm Bureau: Family, Community, and Professionals, 1914–1918," *Agricultural History* 75, no. 4 (2001): 438–69, examines the Farm Bureau's local base in New York, Iowa, and Illinois.

7. E. R. Minns, "Broome County: An Account of Its Agriculture and of Its Farm Bureau," Circular no. 2, Farm Bureau of New York State, 1914, 19, 18.

8. Ibid., 20.

9. New York State Farm Bureau Federation, "The Farm Bureau Carries On: The Growth of an Idea," Ithaca, N.Y., ca. 1945, 5.

10. Minns, "Broome County," 20.

11. New York State Farm Bureau Federation, "Farm Bureau Carries On," 4.

12. See Laurie Winn Carlson, *William J. Spillman and the Birth of Agricultural Economics* (Columbia: University of Missouri Press, 2005).

13. Minns, "Broome County," 20.

14. Ibid.

15. Ibid., 21.

16. See John D. Hicks, *The Populist Revolt: A History of the Farmers' Alliance and the Populist Movement* (Minneapolis: University of Minnesota Press, 1931); Lawrence Goodwyn, *Democratic Promise: The Populist Movement in America* (New York: Oxford University Press, 1976); Lawrence Goodwyn, *The Populist Moment: A Short History of the Agrarian Revolt in America* (New York: Oxford University Press, 1978). For a political science perspective that bridges the late nineteenth and early twentieth centuries, see Elizabeth Sanders, *Roots of Reform: Farmers, Workers, and the American State, 1877–1917* (Chicago: University of Chicago Press, 1999).

17. New York State Farm Bureau Federation, "Farm Bureau Carries On," 7–8.

18. Ibid., 8.

19. Minns, "Broome County," 21.

20. Ibid., 22.

21. Minutes, Farm Improvement Association of Broome County, May 5, 1914. The records of the Farm Improvement Association (hereafter FIA) and its successors, the Farm Bureau Association (hereafter FBA) and the Farm and Home Bureau Association (hereafter FHBA), are in the Broome County Cooperative Extension Archives, Binghamton, N.Y.

22. Minutes, FIA, October 13, 1914.

23. See Maurice Chase Burritt, *The County Agent and the Farm Bureau* (New York: Harcourt, Brace & Co., 1922).

24. New York State Farm Bureau Federation, "Farm Bureau Carries On," 8; minutes, FIA, October 13, 1914; minutes, FHBA, May 21, 1921.

25. Minutes, FIA, December 13, 1914.

26. Minns, "Broome County," 23.

27. Minutes, FIA, August 1915.

28. Minns, "Broome County," 22–23.

29. The survey culminated in Edward G. Misner, "An Economic Study of Dairying on 149 Farms in Broome County, New York," Cornell University Agricultural Experiment Station Bulletin no. 409, Cornell University, April 1922, 273–443.

30. Maurice C. Burritt, "Preface," to Minns, "Broome County," 10.

31. Minns, "Broome County," 11, 14, 16–17.

32. Ibid., 11, 17–19.

33. Ibid., 18, 19, 17. For similar developments in the Midwest, see Mary Neth, "Defining the Rural Problem: Social Policy and Agricultural Institutions," in *Preserving the Family Farm*, 97–121. For the long-term results in the dairy industry, see Hal S. Barron, "Bringing Forth Strife: The Ironies of Dairy Organization in the New York Milkshed," in *Mixed Harvest: The Second Great Transformation in the Rural North, 1870–1930* (Chapel Hill: University of North Carolina Press, 1997), 81–106.

34. Minns, "Broome County," 19.

35. Interview with Richard Rozelle, Eldon and Louella's son.

36. The Grange was already involved in cooperative purchasing. Members purchased lime, phosphate fertilizer, coal, and grass seed in bulk and unloaded it from the railroad cars themselves.

37. The previous year George Young had served as overseer of the Pomona Grange, and Alice Young served as lecturer; minutes, FIA, December 10, 1915.

38. Interview with Richard Rozelle.

39. The Eastman brothers grew up in Berkshire, Tioga County, adjacent to the Nanticoke Valley. Jasper Fay Eastman became a soil chemist. Edward Roe Eastman edited the *Dairymen's League News* from 1917 to 1922 and then became the editor of the *American Agriculturist*.

40. Minutes, FIA, January 2, 1917; February 5, 1917.

41. *Broome County Farm Bureau News* [hereafter *News*], February 1919, 2; August 1919, 2; September 1919, 3; June 1919, 1; minutes, FBA, June 3, 1920. A complete run of the monthly newspaper (later called the *Farm and Home Bureau News*) is in the Broome County Cooperative Extension Archives, Binghamton, N.Y.

42. *News,* January 1919, 2.

43. For the role played by the Grange in establishing the Dairymen's League and the Grange League Federation, see Allen, *History of the New York State Grange,* 80–82, 92–94, 126–31.

44. On the formation of the Home Bureau during World War I through the cooperation of women's organizations in the city as well as the country, see Eastman, "Agriculture," 144–49.

45. Minutes, FIA, December 4, 1917.

46. "A Home Economics Department for Broome County," undated and unsigned typescript (probably written and presented by Ruby Green Smith), in box "Home Bureau History," Broome County Cooperative Extension Archives, Binghamton, N.Y.

47. Ruby Green Smith, *The People's College: History of the New York State Extension Service in Cornell University and the State, 1876–1948* (Ithaca: Cornell University Press, 1949), 139.

48. Minutes, FIA, December 13, 1918; January 25, 1919.

49. Smith, *People's College,* 138–39.

50. Minutes, FHBA, December 12, 1919.

51. Minutes, FIA, January 25, 1919.

52. Minutes, FIA, December 2, 1919. In the 1920 budget, the Farm Department received $5,871 and the Home Department $2,113. Jasper Eastman received a salary of $1,550. Alice Ambler, the Home Bureau agent, was paid only $400 because it was assumed she worked part-time.

53. *The Nation,* June 1, 1918; reprinted in *The Nation,* March 7, 2002, www.thenation.com/doc/19180601/whitehouse (accessed May 8, 2010).

54. Lot J. Emerson, Duane Barnes, Louis Ketchum, and George Young participated in county meetings. The "honor roll of ten-year members" also lists Clement G. Bowers of Maine and W. Carley and Son of Glen Aubrey; *News,* August 1926, 5.

55. Membership list in the back of FIA minutes, dated 1918–1919.

56. Minutes, FHBA, December 1925; December 13, 1926.

57. Interview with Richard Rozelle.

58. Other Broome County correspondents included several women, whose sex is indicated by the "Mrs." that precedes their initials but who are otherwise unidentifiable.

59. *News,* July 1920, 1; August 1920, 1; December 1922, 4; August 1922; minutes, FHBA, June 6, 1922.

60. *News,* October 1920, 1; November 1920, 2; February 1921, 8.

61. On the Ku Klux Klan and its anti-immigrant, anti-Catholic ideology in the early twentieth century, see Kenneth T. Jackson, *The Ku Klux Klan in the City, 1915–1930* (1967; New York: Oxford University Press, 1992). See also Rory M. McVeigh, *The Rise of the Ku Klux*

Klan: Right-Wing Movements and National Politics (Minneapolis: University of Minnesota Press, 2009).

62. "Americanization," editorial, *American Agriculturist,* March 5, 1921.

63. Editorial, *American Agriculturist,* March 26, 1923, 452.

64. Minutes, FHBA, November 1, 1921.

65. Minutes, FHBA, August 28, 1923; December 6, 1923; December 12, 1924.

66. Edward R. Eastman, *The Trouble Maker* (New York: Macmillan, 1925), 32.

CHAPTER 10

1. On the inapplicability of "separate spheres" to women in the Nanticoke Valley, see Nancy Grey Osterud, *Bonds of Community: The Lives of Farm Women in Nineteenth-Century New York* (Ithaca: Cornell University Press, 1991), 4–9, 275–80. For the origins of this ideology, see Nancy F. Cott, *Bonds of Womanhood: "Woman's Sphere" in New England, 1780–1835* (New Haven, Conn.: Yale University Press, 1977).

2. Marilyn Holt, *Linoleum, Better Babies, and the Modern Farm Woman, 1890–1930* (Albuquerque: University of New Mexico Press, 1995).

3. Gould P. Coleman, *Education and Agriculture: A History of the New York State College of Agriculture at Cornell University* (Ithaca: Cornell University Press, 1962), although acknowledging the capitalist orientation of experts' advice to farmers, fails to recognize its gendered character. For a corrective, see Gould Coleman and Sarah Elbert, "Farming Families: The Farm Needs Everyone," *Research in Rural Sociology and Development* 1 (1984): 61–78.

4. For a discussion of this dynamic as it was played out across the state, see Kathleen R. Babbitt, "Production and Consumption in the Countryside: Rural Women and Cooperative Extension Service Home Economists, New York State, 1870–1935," doctoral diss., State University of New York at Binghamton, 1995.

5. "Some Things Farm Women Want," *American Agriculturist,* September 4, 1926, 164.

6. On the history of home economics in New York, see Nancy K. Berlage, "The Establishment of an Applied Social Science: Home Economists, Science, and Reform at Cornell University, 1870–1930," in *Gender and American Social Science: The Formative Years,* ed. Helene Silverberg, 185–231 (Princeton, N.J.: Princeton University Press, 1998); Flora Rose, Esther Stocks, and Michael Whittier, *A Growing College: Home Economics at Cornell* (Ithaca: Cornell University Press, 1969); Sarah Stage and Virginia B. Vincenti, eds., *Rethinking Home Economics: Women and the History of a Profession* (Ithaca: Cornell University Press, 1997).

7. See Martha Van Rensselaer, "Saving Steps," Cornell Reading Course for Farmers' Wives 1, 1901 (reprinted Ithaca: Cornell University, 2000). All these bulletins are compiled in New York State College of Agriculture, *The Cornell Reading Course for the Farm Home* (Ithaca: The College, 1915). For a fascinating analysis of farm women's letters to Van Rensselaer, see Sarah Elbert, "Women and Farming: Changing Structures, Changing Roles," in *Women and Farming: Changing Roles, Changing Structures,* ed. Wava G. Haney and Jane B. Knowles (Boulder, Colo.: Westview Press, 1988), 245–64.

8. Broome County Home Bureau Programs for 1919 and 1920. All Home Bureau documents are in the Broome County Cooperative Extension Archives, Binghamton, New York. For an analysis of rural women's activism on behalf of suffrage, see Paula Baker, *The Moral Frameworks of Public Life: Gender, Politics, and the State in Rural New York, 1870–1930* (New York: Oxford University Press, 1991).

9. Editorial, Home Bureau section, *Broome County Farm Bureau News* [hereafter *News*], March 1920, 8.

10. Home Bureau section, *News,* November 1919, 6.

11. Broome County Home Bureau Program for 1921.

12. Martha Van Rensselaer, 1929, quoted in Ruby Green Smith, *The People's College: History of the New York State Extension Service in Cornell University and the State, 1876–1948* (Ithaca: Cornell University Press, 1949), 156.

13. On social feminism and the Progressive-era notion of "public housekeeping," see Nancy F. Cott, "What's in a Name?: The Limits of 'Social Feminism'; or, Expanding the Vocabulary of Women's History," *Journal of American History* 76 (1989): 808–29; J. Stanley Lemons, *The Woman Citizen: Social Feminism in the 1920s* (Urbana: University of Illinois Press, 1973); Robyn Muncy, *Creating a Female Dominion in American Reform, 1890–1920* (Oxford: Oxford University Press, 1991).

14. Broome County Home Bureau Program for 1920.

15. Broome County Home Bureau Program for 1923.

16. For a discussion of the conflicts between home demonstration agents and rural women over arts and crafts programs, see Kathleen R. Babbitt, "The Productive Farm Woman and the Extension Home Economist in New York State, 1920–1940," *Agricultural History* 67 (spring 1993): 83–101.

17. Broome County Home Bureau Program for 1925–1926.

18. For critiques of this agenda as it was applied to urban women's work, see Susan Strasser, *Never Done: A History of American Housework,* 2nd ed. (New York: Owl Books/Henry Holt, 2000); Ruth Schwartz Cowan, *More Work for Mother: The Ironies of Household Technology from the Open Hearth to the Microwave* (New York: Basic Books, 1983).

19. Broome County Home Bureau Program for 1925–1926.

20. Katherine Jellison, *Entitled to Power: Farm Women and Technology, 1919–1963* (Chapel Hill: University of North Carolina Press, 1993), esp. 10–32.

21. See Virginia Scharff, *Taking the Wheel: Women and the Coming of the Motor Age* (New York: Free Press, 1991).

22. This approach is reminiscent of household advice books published for urban middle-class women in the mid-nineteenth century, which attempted to reassure women of the supreme importance of their domestic role at the same time that they were being stripped of their productive functions. See, for example, Catharine E. Beecher and Harriet Beecher Stowe, *The American Woman's Home* (1869; reprinted New Brunswick, N.J.: Rutgers University Press, 2002).

23. Broome County Home Bureau Program for 1924–1925.

24. Broome County Home Bureau Program for 1923.

25. Ruby Green Smith, "Community Life and the Home Bureau," n.d., in box "Home Bureau History."

26. Ibid., 2.

27. Home Bureau section, *News,* December 1919, 6.

28. Ibid., April 1920, 7; January 1921, 7.

29. Ibid., May 1923, 7; December 1923, 6.

30. Although complete membership lists for the Glen Aubrey, Maine, and Union Center groups have not been preserved, the names of 53 women who hosted meetings, acted as local leaders for specific projects, and served on community committees were published in the *News* or appeared in the annual reports of the Broome County Home Bureau between 1919 and 1928. This analysis is based on those names, which represent a significant proportion of Home Bureau members.

31. See Karen Blair, *The Clubwoman as Feminist: True Womanhood Redefined, 1868–1914* (New York: Holmes and Meier, 1980).

32. *News,* July 1920, 8.

33. Couples who were particularly active include Mr. and Mrs. Duane Barnes and Mr. and Mrs. Nathan Barnes of Glen Aubrey; Mr. and Mrs. Erford Lamb of Nanticoke; Mr. and Mrs. Gilbert Allen and Mr. and Mrs. Henry Ingalls of Maine; Mr. and Mrs. Will Kolb and Mr. and Mrs. George Kolb of West Chenango; Mr. and Mrs. Louis Ketchum, Mr. and Mrs. Harry Woodward, Mr. and Mrs. Wayne Woodward, and Mr. and Mrs. George Young of Union Center.

34. Home Bureau section, *News,* April 1919, 6.

35. Ibid., September 1919, 6.

36. Ibid., June 1920, 6.

37. On the Ladies' Aid Societies in the Protestant churches, see Osterud, *Bonds of Community,* 267–74. The Maine Home Bureau rotated its meetings among the three churches in the village.

38. For farm women's selective adaptation of the recommendations made by home economics extension programs in other regions of the United States, see Mary Neth, *Preserving the Family Farm: Women, Community, and the Foundations of Agribusiness in the Midwest, 1900–1940* (Baltimore: Johns Hopkins University Press, 1995), 214–43; Lu Ann Jones, *Mama Learned Us to Work: Farm Women in the New South* (Chapel Hill: University of North Carolina Press, 2002), 14–22, 107–38. The major collection of oral histories is Eleanor Arnold, ed., *Voices of American Homemakers* (Bloomington: Indiana University Press, 1993).

39. Edward R. Eastman, "Should Women Help with Farm Work? Their Sacrifices, Sometimes Necessary, Have Not Always Paid," *American Agriculturist,* August 30, 1924, 131, 143.

40. "Letters from Our Women Readers: They Say Outdoor Work Does Not Make Women Drudges," *American Agriculturist,* January 30, 1926, 108.

41. *American Agriculturist,* July 3, 1926, 5.

42. *News,* August 1926, 8; November 1926, 8.

43. Playbill in box "Minutes of the Broome County Farm and Home Bureau Association, Dec. 1919–Jan. 1930."

44. In November 1926, Ralph Young noted in his diary, "To Grange Fair at night. Entertainment. Resolved, that girls wearing long hair and long skits make better wives than those with short hair and short skirts. Affirmative H. Woodward and Jim Oakes, Negative Vivian Bradbury and myself." He did not record the outcome of the debate about the flapper. For a literary view of the image of rural women in relation to modernity, see Megan J. Elias, *Stir It Up: Home Economics in American Culture* (Philadelphia: University of Pennsylvania Press, 2008).

45. "Broome County Historical Tour," held July 13, 1928, in box "Home Bureau History."

46. Broome County Home Bureau Suggested New Projects for 1929–1930, in box "Home Bureau History."

47. The Dairymen's League sponsored an essay contest in 1924. In 1927, the county leader of the Home Bureau solicited letters in response to the threatened cutoff of public funds for the support of the program. All the letters received in this essay contest are in folder "Old History," box "Home Bureau History."

48. Home Bureau section, *News*, August 1919, 7.

49. Ibid., March 1921, 6.

50. Ibid., September 1926–1927.

51. Ibid., 7.

52. *Extension Service News*, June 1919.

53. *News*, April 1924, 12.

54. Ibid., March 1924, 12.

55. Ibid., July 1923, 8.

56. Home Bureau section, *News*, November 1920, 6.

57. Farm Bureau section, *News*, September 1921, 12. The next month, members of the Maine Farm and Home Bureau watched a movie on water systems.

58. *News*, May 1923, 11.

59. Editorial, Home Bureau section, *News*, September 1920, 4.

60. Home Bureau section, *News*, 6–7.

61. Editorial, Home Bureau section, *News*, November 1921, 7.

62. *News*, October 1921, 4.

63. Only if a family hired servants did it count; then it figured as a wage to the employee and as an expense to the employer. For a critical analysis of the devaluation of household labor in the political economy of capitalism, see Jeanne Boydston, *Home and Work: Housework, Wages, and the Ideology of Labor in the Early Republic* (Oxford: Oxford University Press, 1990).

64. *News*, July 1924, 7.

65. Ibid., February 1923, 7.

66. Home Bureau section, *News*, July 1920, 6.

67. Ibid., November 1920, 7.

68. "Do It By Machinery," Home Bureau section, *News*, March 1921, 6.

69. Emily Hoag, a rural sociologist at the USDA, analyzed this survey, along with letters written in response to an article in *Farm and Home* on "The Woman God Forgot" and a contest in the *Farmer's Wife* on the question, "Would you like your daughter to marry a farmer?" as well as her own interviews. Her unpublished 1923 report, "The Advantages of Farm Life," emphasizes farm women's pride in their outdoor as well as indoor labor and their integral role in farm operations, in contrast to outsiders' perception of them as downtrodden drudges. See Neth, *Preserving the Family Farm*, 237–38; Jellison, *Entitled to Power*, 27–30.

70. A survey by Nancy Kritser of the Cornell Home Economics Department on how rural women made money at home was announced in Home Bureau section, *News*, November 1926, 7.

71. Farm Bureau proposal, quoted in *News*, July 1923, 3.

72. Home Bureau section, *News*, February 1920, 6.

73. *News*, July 1926, 6.

74. Ibid., 8.

75. Ibid., June 1920, 6. See Edith M. Fox, "Martha Van Rensselaer," in *Notable American Women*, ed. Edward T. James, Janet Wilson James, and Paul S. Boyer, vol. 2, 513–14 (Cambridge, Mass.: Harvard University Press, 1971).

CONCLUSION

1. Franklin Delano Roosevelt, "Second Inaugural Address," January 20, 1937, in *Public Papers and Addresses of Franklin D. Roosevelt, 1928–1945,* available from the American

Presidency Project, University of California at Santa Barbara, www.presidency.ucsb.edu/ws/index.php?pid=15349 (accessed December 19, 2010).

2. R. D. Cooper, "Origin and Development of the Dairymen's League," manuscript, ca. 1938, 91, in the Guide to the Dairymen's League Cooperative Association Records, Division of Rare and Manuscript Collections, Cornell University Library, Ithaca.

3. Ibid., 95.

4. Cooper's history presents the 1933 milk strike as organized by left-wing radicals who were opposed to the Dairymen's League's decision to cooperate with the State Milk Control Board. Cooper, a member of the increasingly conservative leadership of the cooperative organization, describes the strike as fomented by outsiders: "Professional agitators, who had engineered trouble in labor circles and industry generally, were at work among farmers, obtaining the services of local farmers, who are leaders of forces of discontent, and using them as tools of a widespread communistic movement; that, in several sections of the country, advocates of violence had already obstructed and retarded the work of farmers' cooperative marketing associations by fomenting farm holidays, farm strikes and milk strikes, resulting in violence and bloodshed and had even resisted due process of law by attempts to prevent officers of the law from performing their sworn duties." Ibid., 115. The expression of such sentiments in 1938 indicates the gap between the top leadership of the Dairymen's League and the radical farmers' movements taking action in the Midwest and organizing in western New York state, as well as mistrust of the labor movement then being led by the Congress of Industrial Organizations (CIO). The *American Agriculturist,* too, denounced the 1933 strike as led by Jewish communists from New York City who conspired to dupe frustrated farmers into acting like militant labor unionists. For an objective historical analysis, see Lowell K. Dyson, "The Milk Strike of 1939 and the Destruction of the Dairy Farmers' Union," *New York History* 51, no. 5 (October 1970): 523–43. For an account by a leading organizer, see Thomas J. Kriger, "The 1939 Dairy Farmers Union Milk Strike in Heuvelton and Canton, New York: The Story in Words and Pictures," *Journal for MultiMedia History* 1, no. 1 (fall 1998), www.albany.edu/jmmh/vol1no1/dairy.html (accessed December 23, 2010). Few farmers in the Nanticoke Valley participated in the 1939 strike led by the Dairy Farmers' Union. For an analysis of women's roles on both sides of this conflict, see Linda G. Ford, "Another Double Burden: Farm Women and Agrarian Activism in Depression Era New York State," *New York History* (October 1994): 373–98.

5. In industrial workplaces, a *wildcat strike* is conducted by workers without the approval of their union, which has a signed contract with the employer. When workers are divided and some try to go to work while others are on strike, the strikers may try to block the "scabs" from entering the plant. In the 1930s, as before, some conflicts among workers became violent, especially when the police intervened to protect those who tried to cross the picket line. These words were familiar to most Americans when labor unions were organizing in mass-production industries across the country.

6. In the western and north-central parts of the state, this strike pitted smaller-scale dairy farmers who sold milk mostly in the summer, much of which was manufactured into products such as butter and cheese, against larger-scale operators who shipped fluid milk throughout the year. All suffered from low prices, and many resented state action that appeared to threaten the power of the farmers' cooperative. For details on the controversy over the close business relationship that the Dairymen's League had with Borden's, the largest milk processor, see John J. Dillon, *Seven Decades of Milk: A History of New York's Dairy Industry* (New York: Orange Judd Publishing, 1941).

7. For similar patterns in the South, see Melissa Walker, ed., *Country Women Cope with Hard Times: A Collection of Oral Histories* (Columbia: University of South Carolina Press, 2004). On the Plains, see Deborah Fink, "Sidelines and Moral Capital: Women on Nebraska Farms in the 1930s," in *Women and Farming: Changing Roles, Changing Structures,* ed. Wava G. Haney and Jane B. Knowles, 55–72 (Boulder, Colo.: Westview Press, 1988).

8. On the history of making clothing from flour and feed bags, see Lu Ann Jones, "From Feed Bags to Fashion," in *Mama Learned Us to Work: Farm Women in the New South* (Chapel Hill: University of North Carolina Press, 2002), 171–83.

9. Broome County Home Bureau, Minutes of Annual Meeting, 1934, 3, files 29–39, "Minutes—Annual Meetings," in box "Home Bureau History," Broome County Cooperative Extension Archives, Binghamton, N.Y.

10. Ibid, 2.

11. Historians have found similar patterns in other rural regions, although class stratification was visible earlier in places that depended primarily on grain and livestock rather than dairying and poultry and were located far from urban employment and marketing opportunities.

Still, farm capital was depleted during the 1930s, and World War II accelerated changes that swept many farmers off the land. See Mary Neth, *Preserving the Family Farm: Women, Community, and the Foundations of Agribusiness in the Midwest, 1900–1940* (Baltimore: Johns Hopkins University Press, 1995); Jane H. Adams, *The Transformation of Rural Life: Southern Illinois, 1890–1990* (Chapel Hill: University of North Carolina Press, 1994).

12. The consolidation and industrialization of the poultry industry affected rural women across the country. See Deborah Fink, *Open Country, Iowa: Rural Women, Tradition and Change* (Albany: SUNY Press, 1986), 135–59.

13. Indeed, farm succession remained a concern throughout the 1930s. In "Notes from the Southern Tier," E. L. Vincent observed that "Andy Hogg has taken unto himself a bride, but will keep on with his father, Walter Hogg, in general farming in West Chenango, Broome Co. Walter is one of the farmers fortunate enough to keep his boy with him on the home place where Walter's father used to live." *American Agriculturist,* May 3, 1930, 487.

14. Douglas Harper, *Changing Works: Visions of a Lost Agriculture* (Chicago: University of Chicago Press, 2001), uses photographs from a documentary project directed by Roy Stryker for Standard Oil of New Jersey and oral history interviews with farmers in northern New York state to document the social meaning of changing technologies.

15. For a poignant portrait of families living in a straggling hamlet located northwest of the Nanticoke Valley nearer Ithaca, see Janet M. Fitchen, *Poverty in Rural America: A Case Study* (Boulder, Colo.: Westview Press, 1981).

16. What historians call producerist ideology emerged in the early nineteenth century among Jacksonian Democrats and reached its apogee in the late-nineteenth-century Populist Party. On the Populists' roots in farmers' as well as workers' organizations, see Lawrence Goodwyn, *Democratic Promise: The Populist Movement in America* (New York: Oxford University Press, 1976). Producerist ideology and populism (with a small p) cannot be stably located on a left-right political spectrum, however; by the turn of the twentieth century, some spokesmen attacked not only "parasitic" finance capital but also African Americans, Chinese immigrants, and the urban poor. Michael Kazin, *The Populist Persuasion: An American History,* rev. ed. (Ithaca: Cornell University Press, 1998), discusses conservative populist leaders of the 1930s and the definitive shift to the right that occurred after World War II.

17. The Civil War remained salient in public memory. A very high proportion of local men had served in the Union Army, and veterans' organizations held frequent commemorations and participated in all patriotic celebrations. See Nancy Grey Osterud, "Rural Women during the Civil War: New York's Nanticoke Valley, 1861–1865," *New York History* 62, no. 4 (October 1990): 357–85.

18. Lu Ann Jones, who interviewed black and white farmers for the Oral History of Southern Agriculture project conducted by the Smithsonian Institution, found this word appearing, seemingly by itself, on her computer screen as she transcribed interviews that used the phrases "farm family" and "family farm" frequently and interchangeably.

19. See Mary Neth, "Building the Base: Farm Women, the Rural Community, and Farm Organizations in the Midwest, 1900–1940," in *Women and Farming: Changing Roles, Changing Structures,* ed. Wava G. Haney and Jane B. Knowles, 339–55 (Boulder, Colo.: Westview Press, 1988).

20. In addition to the standard histories of the Farm Bureau and the Dairymen's League, see Hal S. Barron, "Bringing Forth Strife: The Ironies of Dairy Organization in the New York Milkshed," in *Mixed Harvest: The Second Great Transformation in the Rural North, 1870–1930* (Chapel Hill: University of North Carolina Press, 1998), 81–105.

21. Many rural organizations that drew whole families into membership also had separate subgroups for adult women, adult men, young people, and children. These age-specific groups did not tend to fracture the larger organizations but, rather, to extend and revitalize them, drawing more people into a lively round of activities and intensifying people's adhesion to the organization. Parallel and integrative structures worked in synchrony.

22. The importance of strong networks of kinship and friendship to the quality of rural women's lives is especially clear when places where they were present are compared to others where they were absent, as Deborah Fink has done. In Iowa, ties among women were crucial to their well-being; see Fink, *Open Country, Iowa,* esp. 77–101. In Nebraska, where farmsteads were scattered and many families moved frequently, women suffered profoundly from social isolation and the absence of mutual aid networks; see Deborah Fink, *Agrarian Women: Wives and Mothers in Rural Nebraska, 1880–1940* (Chapel Hill: University of North Carolina Press, 1992).

23. Karen V. Hansen, *Encounter on the Great Plains: Scandinavian Immigrants and Dakota Indians, 1890–1930* (New York: Oxford University Press, forthcoming).

24. Neth, "Building the Base."

Index

Note: Names marked with an asterisk are pseudonyms.